LONDON MATHEMATICAL SOCIETY LECTURE NOTE SERIES

Managing Editor: Professor N.J. Hitchin, Mathematical Institute,
University of Oxford, 24–29 St Giles, Oxford OX1 3LB, United Kingdom

The titles below are available from booksellers, or, in case of difficulty, from Cambridge Universi

46 *p*-adic Analysis: a short course on recent work, N. KOBLITZ
59 Applicable differential geometry, M. CRAMPIN & F.A.E. PIRANI
66 Several complex variables and complex manifolds II, M.J. FIELD
86 Topological topics, I.M. JAMES (ed)
87 Surveys in set theory, A.R.D. MATHIAS (ed)
88 FPF ring theory, C. FAITH & S. PAGE
90 Polytopes and symmetry, S.A. ROBERTSON
92 Representations of rings over skew fields, A.H. SCHOFIELD
93 Aspects of topology, I.M. JAMES & E.H. KRONHEIMER (e
96 Diophantine equations over function fields, R.C. MASON
97 Varieties of constructive mathematics, D.S. BRIDGES & F. RIC
99 Methods of differential geometry in algebraic topology, M. KAROU
100 Stopping time techniques for analysts and probabilists, L. EGGHE
104 Elliptic structures on 3-manifolds, C.B. THOMAS
105 A local spectral theory for closed operators, I. ERDELYI & WANG SHENG
107 Compactification of Siegel moduli schemes, C.-L. CHAI
109 Diophantine analysis, J. LOXTON & A. VAN DER POORTEN (eds)
113 Lectures on the asymptotic theory of ideals, D. REES
114 Lectures on Bochner-Riesz means, K.M. DAVIS & Y.-C. CHANG
116 Representations of algebras, P.J. WEBB (ed)
119 Triangulated categories in the representation theory of finite-dimensional algebras, D. HAPPEL
121 Proceedings of *Groups - St Andrews 1985*, E. ROBERTSON & C. CAMPBELL (eds)
128 Descriptive set theory and the structure of sets of uniqueness, A.S. KECHRIS & A. LOUVEAU
130 Model theory and modules, M. PREST
131 Algebraic, extremal & metric combinatorics, M.-M. DEZA, P. FRANKL & I.G. ROSENBERG (eds)
132 Whitehead groups of finite groups, ROBERT OLIVER
138 Analysis at Urbana, II, E. BERKSON, T. PECK, & J. UHL (eds)
139 Advances in homotopy theory, S. SALAMON, B. STEER & W. SUTHERLAND (eds)
140 Geometric aspects of Banach spaces, E.M. PEINADOR & A. RODES (eds)
141 Surveys in combinatorics 1989, J. SIEMONS (ed)
144 Introduction to uniform spaces, I.M. JAMES
146 Cohen-Macaulay modules over Cohen-Macaulay rings, Y. YOSHINO
148 Helices and vector bundles, A.N. RUDAKOV *et al*
149 Solitons, nonlinear evolution equations and inverse scattering, M. ABLOWITZ & P. CLARKSON
150 Geometry of low-dimensional manifolds 1, S. DONALDSON & C.B. THOMAS (eds)
151 Geometry of low-dimensional manifolds 2, S. DONALDSON & C.B. THOMAS (eds)
152 Oligomorphic permutation groups, P. CAMERON
153 L-functions and arithmetic, J. COATES & M.J. TAYLOR (eds)
155 Classification theories of polarized varieties, TAKAO FUJITA
156 Twistors in mathematics and physics, T.N. BAILEY & R.J. BASTON (eds)
158 Geometry of Banach spaces, P.F.X. MÜLLER & W. SCHACHERMAYER (eds)
159 Groups St Andrews 1989 volume 1, C.M. CAMPBELL & E.F. ROBERTSON (eds)
160 Groups St Andrews 1989 volume 2, C.M. CAMPBELL & E.F. ROBERTSON (eds)
161 Lectures on block theory, BURKHARD KÜLSHAMMER
162 Harmonic analysis and representation theory, A. FIGA-TALAMANCA & C. NEBBIA
163 Topics in varieties of group representations, S.M. VOVSI
164 Quasi-symmetric designs, M.S. SHRIKANDE & S.S. SANE
166 Surveys in combinatorics, 1991, A.D. KEEDWELL (ed)
168 Representations of algebras, H. TACHIKAWA & S. BRENNER (eds)
169 Boolean function complexity, M.S. PATERSON (ed)
170 Manifolds with singularities and the Adams-Novikov spectral sequence, B. BOTVINNIK
171 Squares, A.R. RAJWADE
172 Algebraic varieties, GEORGE R. KEMPF
173 Discrete groups and geometry, W.J. HARVEY & C. MACLACHLAN (eds)
174 Lectures on mechanics, J.E. MARSDEN
175 Adams memorial symposium on algebraic topology 1, N. RAY & G. WALKER (eds)
176 Adams memorial symposium on algebraic topology 2, N. RAY & G. WALKER (eds)
177 Applications of categories in computer science, M. FOURMAN, P. JOHNSTONE & A. PITTS (eds)
178 Lower K- and L-theory, A. RANICKI
179 Complex projective geometry, G. ELLINGSRUD *et al*
180 Lectures on ergodic theory and Pesin theory on compact manifolds, M. POLLICOTT
181 Geometric group theory I, G.A. NIBLO & M.A. ROLLER (eds)
182 Geometric group theory II, G.A. NIBLO & M.A. ROLLER (eds)
183 Shintani zeta functions, A. YUKIE
184 Arithmetical functions, W. SCHWARZ & J. SPILKER
185 Representations of solvable groups, O. MANZ & T.R. WOLF
186 Complexity: knots, colourings and counting, D.J.A. WELSH
187 Surveys in combinatorics, 1993, K. WALKER (ed)
188 Local analysis for the odd order theorem, H. BENDER & G. GLAUBERMAN
189 Locally presentable and accessible categories, J. ADAMEK & J. ROSICKY
190 Polynomial invariants of finite groups, D.J. BENSON

191 Finite geometry and combinatorics, F. DE CLERCK *et al*
192 Symplectic geometry, D. SALAMON (ed)
194 Independent random variables and rearrangement invariant spaces, M. BRAVERMAN
195 Arithmetic of blowup algebras, WOLMER VASCONCELOS
196 Microlocal analysis for differential operators, A. GRIGIS & J. SJÖSTRAND
197 Two-dimensional homotopy and combinatorial group theory, C. HOG-ANGELONI *et al*
198 The algebraic characterization of geometric 4-manifolds, J.A. HILLMAN
199 Invariant potential theory in the unit ball of $\mathbf{C}^n$, MANFRED STOLL
200 The Grothendieck theory of dessins d'enfant, L. SCHNEPS (ed)
201 Singularities, JEAN-PAUL BRASSELET (ed)
202 The technique of pseudodifferential operators, H.O. CORDES
203 Hochschild cohomology of von Neumann algebras, A. SINCLAIR & R. SMITH
204 Combinatorial and geometric group theory, A.J. DUNCAN, N.D. GILBERT & J. HOWIE (eds)
205 Ergodic theory and its connections with harmonic analysis, K. PETERSEN & I. SALAMA (eds)
207 Groups of Lie type and their geometries, W.M. KANTOR & L. DI MARTINO (eds)
208 Vector bundles in algebraic geometry, N.J. HITCHIN, P. NEWSTEAD & W.M. OXBURY (eds)
209 Arithmetic of diagonal hypersurfaces over finite fields, F.Q. GOUVÊA & N. YUI
210 Hilbert C*-modules, E.C. LANCE
211 Groups 93 Galway / St Andrews I, C.M. CAMPBELL *et al* (eds)
212 Groups 93 Galway / St Andrews II, C.M. CAMPBELL *et al* (eds)
214 Generalised Euler-Jacobi inversion formula and asymptotics beyond all orders, V. KOWALENKO *et al*
215 Number theory 1992–93, S. DAVID (ed)
216 Stochastic partial differential equations, A. ETHERIDGE (ed)
217 Quadratic forms with applications to algebraic geometry and topology, A. PFISTER
218 Surveys in combinatorics, 1995, PETER ROWLINSON (ed)
220 Algebraic set theory, A. JOYAL & I. MOERDIJK
221 Harmonic approximation, S.J. GARDINER
222 Advances in linear logic, J.-Y. GIRARD, Y. LAFONT & L. REGNIER (eds)
223 Analytic semigroups and semilinear initial boundary value problems, KAZUAKI TAIRA
224 Computability, enumerability, unsolvability, S.B. COOPER, T.A. SLAMAN & S.S. WAINER (eds)
225 A mathematical introduction to string theory, S. ALBEVERIO, J. JOST, S. PAYCHA, S. SCARLATTI
226 Novikov conjectures, index theorems and rigidity I, S. FERRY, A. RANICKI & J. ROSENBERG (eds)
227 Novikov conjectures, index theorems and rigidity II, S. FERRY, A. RANICKI & J. ROSENBERG (eds)
228 Ergodic theory of $\mathbf{Z}^d$ actions, M. POLLICOTT & K. SCHMIDT (eds)
229 Ergodicity for infinite dimensional systems, G. DA PRATO & J. ZABCZYK
230 Prolegomena to a middlebrow arithmetic of curves of genus 2, J.W.S. CASSELS & E.V. FLYNN
231 Semigroup theory and its applications, K.H. HOFMANN & M.W. MISLOVE (eds)
232 The descriptive set theory of Polish group actions, H. BECKER & A.S. KECHRIS
233 Finite fields and applications, S. COHEN & H. NIEDERREITER (eds)
234 Introduction to subfactors, V. JONES & V.S. SUNDER
235 Number theory 1993–94, S. DAVID (ed)
236 The James forest, H. FETTER & B. GAMBOA DE BUEN
237 Sieve methods, exponential sums, and their applications in number theory, G.R.H. GREAVES *et al*
238 Representation theory and algebraic geometry, A. MARTSINKOVSKY & G. TODOROV (eds)
239 Clifford algebras and spinors, P. LOUNESTO
240 Stable groups, FRANK O. WAGNER
241 Surveys in combinatorics, 1997, R.A. BAILEY (ed)
242 Geometric Galois actions I, L. SCHNEPS & P. LOCHAK (eds)
243 Geometric Galois actions II, L. SCHNEPS & P. LOCHAK (eds)
244 Model theory of groups and automorphism groups, D. EVANS (ed)
245 Geometry, combinatorial designs and related structures, J.W.P. HIRSCHFELD *et al*
246 p-Automorphisms of finite p-groups, E.I. KHUKHRO
247 Analytic number theory, Y. MOTOHASHI (ed)
248 Tame topology and o-minimal structures, LOU VAN DEN DRIES
249 The atlas of finite groups: ten years on, ROBERT CURTIS & ROBERT WILSON (eds)
250 Characters and blocks of finite groups, G. NAVARRO
251 Gröbner bases and applications, B. BUCHBERGER & F. WINKLER (eds)
252 Geometry and cohomology in group theory, P. KROPHOLLER, G. NIBLO, R. STÖHR (eds)
253 The q-Schur algebra, S. DONKIN
254 Galois representations in arithmetic algebraic geometry, A.J. SCHOLL & R.L. TAYLOR (eds)
255 Symmetries and integrability of difference equations, P.A. CLARKSON & F.W. NIJHOFF (eds)
256 Aspects of Galois theory, HELMUT VÖLKLEIN *et al*
257 An introduction to noncommutative differential geometry and its physical applications 2ed, J. MADORE
258 Sets and proofs, S.B. COOPER & J. TRUSS (eds)
259 Models and computability, S.B. COOPER & J. TRUSS (eds)
260 Groups St Andrews 1997 in Bath, I, C.M. CAMPBELL *et al*
261 Groups St Andrews 1997 in Bath, II, C.M. CAMPBELL *et al*
263 Singularity theory, BILL BRUCE & DAVID MOND (eds)
264 New trends in algebraic geometry, K. HULEK, F. CATANESE, C. PETERS & M. REID (eds)
265 Elliptic curves in cryptography, I. BLAKE, G. SEROUSSI & N. SMART
267 Surveys in combinatorics, 1999, J.D. LAMB & D.A. PREECE (eds)
268 Spectral asymptotics in the semi-classical limit, M. DIMASSI & J. SJÖSTRAND
269 Ergodic theory and topological dynamics, M. B. BEKKA & M. MAYER
270 Analysis on Lie Groups, N. T. VAROPOULOS & S. MUSTAPHA
271 Singular perturbations of differential operators, S. ALBERVERIO & P. KURASOV
272 Character theory for the odd order function, T. PETERFALVI
273 Spectral theory and geometry, E. B. DAVIES & Y. SAFAROV (eds)
274 The Mandlebrot set: theme and variation, TAN LEI (ed)

London Mathematical Society Lecture Note Series. 265

Elliptic Curves in Cryptography

I. F. Blake
Hewlett-Packard Laboratories, Palo Alto

G. Seroussi
Hewlett-Packard Laboratories, Palo Alto

N. P. Smart
Hewlett-Packard Laboratories, Bristol

PUBLISHED BY THE PRESS SYNDICATE OF THE UNIVERSITY OF CAMBRIDGE
The Pitt Building, Trumpington Street, Cambridge, United Kingdom

CAMBRIDGE UNIVERSITY PRESS
The Edinburgh Building, Cambridge CB2 2RU, UK
40 West 20th Street, New York, NY 10011–4211, USA
477 Williamstown Road, Port Melbourne, VIC 3207, Australia
Ruiz de Alarcón 13, 28014 Madrid, Spain
Dock House, The Waterfront, Cape Town 8001, South Africa

http://www.cambridge.org

First published 1999
Reprinted 2000 (three times), 2001, 2002

Printed in the United Kingdom at the University Press, Cambridge

A catalogue record for this book is available from the British Library

Library of Congress Cataloguing in Publication data
Blake Ian F.
Elliptic Curves in Cryptography / I.F. Blake, G. Seroussi, N.P. Smart
p. cm. – (London Mathematical Society Lecture Note Series; 265)
Includes bibliographical references and index.
ISBN 0 521 65374 6 (pbk.)
1. Computer security. 2. Cryptography. 3. Curves, Elliptic–Data processing.
I. Seroussi, G. (Gadiel), 1955 - II. Smart, Nigel P. (Nigel Paul), 1967 -.
III. Title. IV. Series.
QA76.9.A25.B27 1999
005.8′.2–dc21 99–19696 CIP

ISBN 0 521 65374 6 paperback

To

Elizabeth, Lauren and Michael,

Lidia, Ariel and Dahlia,

Maggie, Ellie and Oliver.

Contents

Preface xi

Abbreviations and Standard Notation xiii

Chapter I. Introduction 1
- I.1. Cryptography Based on Groups 2
- I.2. What Types of Group are Used 6
- I.3. What it Means in Practice 8

Chapter II. Finite Field Arithmetic 11
- II.1. Fields of Odd Characteristic 11
- II.2. Fields of Characteristic Two 19

Chapter III. Arithmetic on an Elliptic Curve 29
- III.1. General Elliptic Curves 30
- III.2. The Group Law 31
- III.3. Elliptic Curves over Finite Fields 34
- III.4. The Division Polynomials 39
- III.5. The Weil Pairing 42
- III.6. Isogenies, Endomorphisms and Torsion 44
- III.7. Various Functions and q-Expansions 46
- III.8. Modular Polynomials and Variants 50

Chapter IV. Efficient Implementation of Elliptic Curves 57
- IV.1. Point Addition 57
- IV.2. Point Multiplication 62
- IV.3. Frobenius Expansions 73
- IV.4. Point Compression 76

Chapter V. The Elliptic Curve Discrete Logarithm Problem 79
- V.1. The Simplification of Pohlig and Hellman 80
- V.2. The MOV Attack 82
- V.3. The Anomalous Attack 88
- V.4. Baby Step/Giant Step 91
- V.5. Methods based on Random Walks 93
- V.6. Index Calculus Methods 97
- V.7. Summary 98

Chapter VI. Determining the Group Order 101
VI.1. Main Approaches 101
VI.2. Checking the Group Order 103
VI.3. The Method of Shanks and Mestre 104
VI.4. Subfield Curves 104
VI.5. Searching for Good Curves 106

Chapter VII. Schoof's Algorithm and Extensions 109
VII.1. Schoof's Algorithm 109
VII.2. Beyond Schoof 114
VII.3. More on the Modular Polynomials 118
VII.4. Finding Factors of Division Polynomials through Isogenies: Odd Characteristic 122
VII.5. Finding Factors of Division Polynomials through Isogenies: Characteristic Two 133
VII.6. Determining the Trace Modulo a Prime Power 138
VII.7. The Elkies Procedure 139
VII.8. The Atkin Procedure 140
VII.9. Combining the Information from Elkies and Atkin Primes 142
VII.10. Examples 144
VII.11. Further Discussion 147

Chapter VIII. Generating Curves using Complex Multiplication 149
VIII.1. The Theory of Complex Multiplication 149
VIII.2. Generating Curves over Large Prime Fields using CM 151
VIII.3. Weber Polynomials 155
VIII.4. Further Discussion 157

Chapter IX. Other Applications of Elliptic Curves 159
IX.1. Factoring Using Elliptic Curves 159
IX.2. The Pocklington–Lehmer Primality Test 162
IX.3. The ECPP Algorithm 164
IX.4. Equivalence between DLP and DHP 166

Chapter X. Hyperelliptic Cryptosystems 171
X.1. Arithmetic of Hyperelliptic Curves 171
X.2. Generating Suitable Curves 173
X.3. The Hyperelliptic Discrete Logarithm Problem 176

Appendix A. Curve Examples 181
A.1. Odd Characteristic 181
A.2. Characteristic Two 186

Bibliography 191

Author Index 199

Subject Index 201

Preface

Much attention has recently been focused on the use of elliptic curves in public key cryptography, first proposed in the work of Koblitz [**62**] and Miller [**103**]. The motivation for this is the fact that there is no known sub-exponential algorithm to solve the discrete logarithm problem on a general elliptic curve. In addition, as will be discussed in Chapter I, the standard protocols in cryptography which make use of the discrete logarithm problem in finite fields, such as Diffie–Hellman key exchange, ElGamal encryption and digital signature, Massey–Omura encryption and the Digital Signature Algorithm (DSA), all have analogues in the elliptic curve case.

Cryptosystems based on elliptic curves are an exciting technology because for the same level of security as systems such as RSA [**134**], using the current knowledge of algorithms in the two cases, they offer the benefits of smaller key sizes and hence of smaller memory and processor requirements. This makes them ideal for use in smart cards and other environments where resources such as storage, time, or power are at a premium.

Some researchers have expressed concern that the basic problem on which elliptic curve systems are based has not been looked at in as much detail as, say, the factoring problem, on which systems such as RSA are based. However, all such systems based on the perceived difficulty of a mathematical problem live in fear of a dramatic breakthrough to some extent, and this issue is not addressed further in this work.

This book discusses various issues surrounding the use of elliptic curves in cryptography, including:

- The basic arithmetic operations, not only on the curves but also over finite fields.
- Ways of efficiently implementing the basic operation of adding a point to itself a large number of times (point multiplication).
- Known attacks on systems based on elliptic curves.
- A large section devoted to computing the number of rational points on elliptic curves over finite fields.
- A discussion on the generalization of elliptic curve systems to hyperelliptic systems.

The book is written for a wide audience ranging from the mathematician who knows about elliptic curves (or has been acquainted with them) and who wants a quick survey of the main results pertaining to cryptography, to an implementer who requires some knowledge of elliptic curve mathematics for use in a practical cryptosystem. Clearly, aiming for such diverse audiences is hard, and not all parts of the book will be of the same level of interest to all readers. However, most of the important points such as implementation issues, security issues and point counting issues can be acquired with only a moderate understanding of the underlying mathematics.

We try and give a flavour of the mathematics involved for those who are interested. We decided however not to include most proofs since that not only would dramatically increase the size of the book but also would not serve its main purpose. It is hoped that the numerous references cited and the extensive bibliography provided will direct the interested reader to appropriate sources for all the missing details. In fact, much of the necessary mathematical background can be found in the books by Silverman, [**147**] and [**148**].

Some of the topics covered in the book by Menezes [**97**] are expanded upon. In particular the improvements made to the algorithm of Schoof [**141**] for determining the number of rational points on an elliptic curve are explained, and the method of finding curves using the theory of complex multiplication is discussed. This latter method has other applications when one uses elliptic curves to construct proofs of primality. We also give the first treatment in book form of such methods as point compression (including x-coordinate compression), the attack on anomalous curves and the generalization of the MOV attack to curves such as those with the trace of Frobenius equal to two. Two chapters are devoted to implementation issues. One covers finite fields while the second covers the various techniques available for point multiplication. In addition, the chapter on Schoof's algorithm and its improvements provides algorithmic summaries intended to facilitate the implementation of these point counting techniques.

We would like to thank D. Boneh, S. Galbraith, A.J. Menezes, K. Paterson, M. Rubinstein, E. Scheafer, R. Schoof and S. Zaba who have looked over various portions of the manuscript and given us their comments. All of the remaining mistakes and problems are our own and we apologize in advance for any you may find. The authors would also like to thank Dan Boneh, Johannes Buchmann, Markus Maurer and Volker Müller for many discussions on elliptic curves, their assistance with the implementation of point counting algorithms and the prompt answering of many queries. Thanks are due also to John Cremona for his LaTeX algorithm template which we modified to produce the algorithms in this book.

Finally thanks are due to Hewlett–Packard Company and our colleagues and managers there for their support, assistance and encouragement during the writing of this book.

Abbreviations and Standard Notation

Abbreviations

The following abbreviations of standard phrases are used throughout the book:

AES	Advanced Encryption Standard
BSGS	baby step/giant step method
CM	Complex multiplication
CRT	Chinese Remainder Theorem
DES	Data Encryption Standard
DHP	Diffie–Hellman problem
DLP	Discrete logarithm problem
DSA	Digital Signature Algorithm
ECDLP	Elliptic curve discrete logarithm problem
ECM	Elliptic curve factoring method
ECPP	Elliptic curve primality proving method
GCD	Greatest common divisor
LCM	Least common multiple
MOV	Menezes–Okamoto–Vanstone attack
NAF	Non-adjacent form
NFS	Number field sieve
ONB	Optimal normal basis
RNS	Residue number system
RSA	Rivest–Shamir–Adleman encryption scheme
SD	Signed digit
SEA	Schoof–Elkies–Atkin algorithm

Standard notation

The following standard notation is used throughout the book, often without further definition. Other notation is defined locally near its first use.

K^*, $K^+, \overline{K}$	for a field K, the multiplicative group, additive group, and algebraic closure, respectively
$\mathrm{Gal}(K/F)$	Galois group of K over F
$\mathrm{Aut}(G)$	Automorphism group of G
$\mathrm{char}(K)$	characteristic of K
$\gcd(f,g), \mathrm{lcm}(f,g)$	GCD, LCM of f and g
$\deg(f)$	degree of a polynomial f
$\mathrm{ord}(g)$	order of an element g in a group
$\mathbb{Z}$, $\mathbb{Q}$, $\mathbb{R}$, $\mathbb{C}$	integers, rationals, reals and complex numbers
$\mathbb{Z}_{>k}$	integers greater than k; similarly for $\geq, <, \leq$
$\mathbb{Z}/n\mathbb{Z}$	integers modulo n
$\mathbb{Z}_p$, $\mathbb{Q}_p$	p-adic integers and numbers, respectively
$\mathbb{F}_q$	finite field with q elements
$\mathrm{Tr}_{q\vert p}(x)$	trace of $x \in \mathbb{F}_q$ over $\mathbb{F}_p$, $q = p^n$
$\langle g \rangle$	cyclic group generated by g
$\#S$	cardinality of the set S
E	elliptic curve (equation)
$E(K)$	group of K-rational points on E
$[m]P$	multiplication-by-m map applied to the point P
$E[m]$	group of m-torsion points on the elliptic curve E
$\mathrm{End}(E)$	Endormorphism ring of E
$\mathcal{O}$	point at infinity (on an elliptic curve)
$\wp$	Weierstrass 'pay' function
φ	Frobenius map
ϕ_{Eul}	Euler totient function
$GL_2(R)$	general linear group over the ring R: 2×2 matrices over R with determinant a unit in R
$PGL_2(K)$	projective general linear group over the field K, with scalar multiples identified
$SL_2(\mathbb{Z})$	special linear group of 2×2 matrices over $\mathbb{Z}$ with determinant one
$\left(\frac{\cdot}{p}\right)$	Legendre symbol
$\mathrm{Re}(z), \mathrm{Im}(z)$	real and imaginary parts of $z \in \mathbb{C}$, respectively
$\mathcal{H}$	Poincaré half-plane $\mathrm{Im}(z) > 0$
$O(f(n))$	function $g(n)$ such that $\vert g(n)\vert \leq c\vert f(n)\vert$ for some constant $c > 0$ and all sufficiently large n
$o(f(n))$	function $g(n)$ such that $\lim_{n\to\infty}(g(n)/f(n)) = 0$
$\log_b x$	logarithm to base b of x; natural log if b omitted

Often we will need to present binary, hexadecimal or decimal numbers which are too long to fit on one line. We shall use the standard convention of breaking the number into multiple lines, with a backslash at the end of a line indicating that the number is continued in the next line. For example

$$\begin{aligned} p &= 2^{230} + 67 \\ &= 1725436586697640946858688965569256363112777243042 5\backslash \\ &\quad 9663879063105594989 1. \end{aligned}$$

CHAPTER I

Introduction

We introduce the three main characters in public key cryptography. As in many books on the subject, it is assumed that Alice and Bob wish to perform some form of communication whilst Eve is an eavesdropper who wishes to spy on (or tamper with) the communications between Alice and Bob. Of course there is no assumption that Alice and Bob (or Eve) are actually human. They may (and probably will) be computers on some network such as the Internet.

Modern cryptography, as applied in the commercial world, is concerned with a number of problems. The most important of these are:

1. **Confidentiality:** A message sent from Alice to Bob cannot be read by anyone else.
2. **Authenticity:** Bob knows that only Alice could have sent the message he has just received.
3. **Integrity:** Bob knows that the message from Alice has not been tampered with in transit.
4. **Non-repudiation:** It is impossible for Alice to turn around later and say she did not send the message.

To see why all four properties are important consider the following scenario. Alice wishes to buy some item over the Internet from Bob. She sends her instruction to Bob which contains her credit card number and payment details. She requires that this communication be confidential, since she wants other people to know neither her credit card details nor what she is buying. Bob needs to know that the message is authentic in that it came from Alice and not some impostor. Both Alice and Bob need to be certain that the message's integrity is preserved, for example the amount cannot be altered by some third party whilst it is in transit. Finally Bob requires the non-repudiation property, meaning that Alice should not be able to say she did not send the instruction.

In other words, we require transactions to take place between two mutually distrusting parties over a public network. This is different from conventional private networks, such as those used in banking, where there are key hierarchies and tamper proof hardware which can store symmetric keys.

It is common in the literature to introduce public key techniques in the area of confidentiality protection. Public key techniques are, however, usually infeasible to use directly in this context, being orders of magnitude slower than symmetric techniques. Their use in confidentiality is often limited to the transmission of symmetric cipher keys. On the other hand *digital signatures*, which give the user the authentication, integrity and non-repudiation properties required in electronic commerce, seem to require the use of public key cryptography.

A computer which is processing payments for a bank or a business may need to verify or create thousands of digital signatures every second. This has led to the demand for public key digital signature schemes which are very efficient. Whilst many schemes are based on the discrete logarithm problem in a finite abelian group, there is some debate as to what type of groups to use. One choice is the group of points on an elliptic curve over a finite field. This choice is becoming increasingly popular, precisely because of efficiency considerations. In this book, we attempt to summarize the latest knowledge available on both theoretical and practical issues related to elliptic curve cryptosystems.

I.1. Cryptography Based on Groups

In this section, some of the standard protocols of public key cryptography are surveyed. A more detailed discussion of all of these protocols and other related areas of cryptography can be found in the books by Menezes, van Oorschot and Vanstone [**99**] and Schneier [**139**], although neither of these books covers the use of elliptic curves in cryptography. The protocols discussed here only require the use of a finite abelian group G, of order $\#G$, which is assumed to be cyclic. The group of interest in this work is the *additive* group of points on an elliptic curve. However, it is convenient for the remainder of this chapter to assume the group is *multiplicative*, with generator g, and that the order, $\#G$, is a prime. If this is not the case, we can always take a prime order subgroup of G as our group, with no loss of security. The additive vs. multiplicative issue is, of course, just one of notation. We will revert to additive notation later on, when the discussion focuses on the elliptic curve groups.

The group G should be presented in such a way as to make multiplication and exponentiation easy, whilst computing discrete logarithms is hard. The reason for this will become clearer below. It should also be possible to generate random elements from the group with an almost uniform distribution.

By the *discrete logarithm problem* (DLP) we mean the problem of determining the least positive integer, x, if it exists, which satisfies the equation

$$h = g^x$$

for two, given, elements h and g in the group G. Note that a common feature of all of the following schemes is that if there is a fast way to solve the DLP in G, then they are all insecure for the group G. Since we have assumed that G is of prime order such a discrete logarithm always exists.

I.1.1. Diffie–Hellman key exchange. Alice and Bob wish to agree on a secret random element in the group, which could be of use as a key for a higher speed symmetric algorithm like the *Data Encryption Standard* (DES). They wish to make this agreement over an insecure channel, without having exchanged any information previously. The only public items, which can be shared amongst a group of users, are the group G and an element $g \in G$ of large known order.

1. Alice generates a random integer $x_A \in \{1, \ldots, \#G - 1\}$. She sends to Bob the element
$$g^{x_A}.$$
2. Bob generates a random integer $x_B \in \{1, \ldots, \#G - 1\}$. He sends to Alice the element
$$g^{x_B}.$$
3. Alice can then compute
$$g^{x_A x_B} = (g^{x_B})^{x_A}.$$
4. Likewise, Bob can compute
$$g^{x_A x_B} = (g^{x_A})^{x_B}.$$

The only information that Eve knows is G, g, g^{x_A} and g^{x_B}. If Eve can recover $g^{x_A x_B}$ from this data then Eve is said to have solved a *Diffie–Hellman problem* (DHP). It is easy to see that if Eve can find discrete logarithms in G then she can solve the DHP. It is believed for most groups in use in cryptography that the DHP and the DLP are equivalent [**94**], in a complexity-theoretic sense (there is a polynomial time reduction of one problem to the other, and vice versa).

I.1.2. ElGamal encryption [**39**]. Alice wishes to send a message to Bob. Her message, m, is assumed to be encoded as an element in the group. Bob has a public key consisting of g and $h = g^x$, where x is the private key.

1. Alice generates a random integer $k \in \{1, \ldots, \#G - 1\}$ and computes
$$a = g^k, \ b = h^k m.$$
2. Alice sends the cipher text (a, b) to Bob.
3. Bob can recover the message from the equation
$$ba^{-x} = h^k m g^{-kx} = g^{xk - xk} m = m.$$

I.1.3. ElGamal digital signature [**39**]. Here, Bob wants to sign a message $m \in (\mathbb{Z}/(\#G)\mathbb{Z})$. He can use the same public and private key pair, h and x, as he used for the encryption scheme. We will need a bijection f from G to $\mathbb{Z}/(\#G)\mathbb{Z}$.

1. Bob generates a random integer $k \in \{1, \ldots, \#G - 1\}$, and computes
$$a = g^k.$$
2. Bob computes a solution, $b \in \mathbb{Z}/(\#G)\mathbb{Z}$, to the congruence
$$m \equiv x f(a) + bk \pmod{\#G}.$$

3. Bob sends the signature, (a, b), and the message, m, to Alice.
4. Alice verifies the signature by checking that the following equation holds:

$$h^{f(a)}a^b = g^{xf(a)+kb} = g^m.$$

I.1.4. Digital Signature Algorithm. A version of ElGamal signatures, called the *Digital Signature Algorithm* (DSA), is the basis of the Digital Signature Standard [**FIPS186**]. An elliptic curve version of DSA (ECDSA) is described in the IEEE P1363 standard draft [**P1363**]. The signature procedure is almost identical to the ElGamal scheme above. It is described here for the sake of completeness, as well as to introduce a slightly different signature verification procedure with some computational advantages.

Bob wants to sign a message $m \in \mathbb{Z}/(\#G)\mathbb{Z}$. He uses the same public and private key pair h and x as before, and both he and Alice use a common bijective mapping, f, from G to $\mathbb{Z}/(\#G)\mathbb{Z}$.

1. Bob generates a random integer $k \in \{1, \ldots, \#G - 1\}$, and computes

$$a = g^k.$$

2. He computes the solution, b, to the congruence

$$m \equiv -xf(a) + kb \pmod{\#G}.$$

3. He sends the signature, (a, b), and the message, m, to Alice.
4. Alice computes

$$u = mb^{-1} \pmod{\#G}\ ,\ v = f(a)b^{-1} \pmod{\#G}.$$

5. She then computes

$$w = g^u h^v$$

and verifies that

$$\begin{aligned} w &= g^u h^v = g^{mb^{-1}} g^{vx} = g^{mb^{-1}+xf(a)b^{-1}} \\ &= g^{(m+xf(a))b^{-1}} = g^{kbb^{-1}} = g^k \\ &= a. \end{aligned}$$

Although the signature verification procedure implemented by Alice appears, at first glance, more complicated than the one described for the ElGamal scheme, it is in fact computationally simpler. Upon closer scrutiny, one notes that the verification procedure described for DSA requires two group exponentiations, while the one described for the ElGamal scheme requires three. The two procedures are, of course, mathematically equivalent.

In its standardized versions, the DSA requires also a secure *hashing function.* This is a many-to-one function that maps the original message to a shorter *digest*, in a way that is infeasible to invert in practice. The message digest is the quantity actually operated on, in lieu of m. See, e.g., [**99**] or [**P1363**] for the details.

I.1.5. Massey–Omura encryption. Here Alice wishes to send a message to Bob. They do not need to have a private or public key. The message is encoded as an element $m \in G$. This protocol is sometimes described as the 'you-to-me, me-to-you' method. It requires Alice and Bob to carry out a conversation rather than just a single transmission of encrypted text.

1. Alice computes a random integer, x_A, coprime to $\#G$, and sends Bob the element
$$a = m^{x_A}.$$
2. Bob computes a random integer, x_B, coprime to $\#G$, and sends back to Alice the element
$$b = a^{x_B} = m^{x_A x_B}.$$
3. Alice can compute $x_A^{-1} \pmod{\#G}$ and so sends back to Bob the element
$$a' = b^{x_A^{-1}} = m^{x_A x_B x_A^{-1}} = m^{x_B}.$$
4. Finally Bob computes $x_B^{-1} \pmod{\#G}$ and can decrypt the message as
$$(a')^{x_B^{-1}} = m^{x_B x_B^{-1}} = m.$$

This algorithm, also referred to as the 'double lock' algorithm, is seldom used in practice but is of historical interest.

I.1.6. Nyberg–Rueppel digital signature [113]. Nyberg and Rueppel present a series of digital signature schemes which allow message recovery. Below we give a variant of one of these schemes, based on a system of Piveteau [**122**]. However, here it is given as a standard signature scheme without any message recovery. For details on how to add message recovery, to this and to other schemes, we refer the reader to [**113**].

Our reason for including the following scheme is that the message to be signed, m, is a member of the group G and not $\mathbb{Z}/(\#G)\mathbb{Z}$. This makes it slightly different from the ElGamal and DSA schemes above.

Once again we assume f is a bijection from G to $\mathbb{Z}/(\#G)\mathbb{Z}$. Alice wishes to sign a message, $m \in G$. She has a private key $x \in \mathbb{Z}$, coprime to $\#G$, and a public key $y = g^x$.

1. She computes a random integer, k, coprime to $\#G$, and computes $r = g^{-k}m$.
2. Alice then computes a solution, s, to the congruence
$$1 \equiv f(r)x + sk \pmod{\#G}.$$
3. She sends the message, m, and the digital signature, (r, s), to Bob.
4. Bob can verify that the message came from Alice by verifying the equation
$$y^{-f(r)} r^s = g^{sk-1-sk} m^s = m^s g^{-1}.$$

I.1.7. Problem reductions. It is not proven that breaking any of the above schemes is equivalent to solving the DLP, but this is believed to be the case. That no proof of this fact has been found is similar to other situations in cryptography: for example there is no proof that breaking the RSA system ([**133**] [**134**]) is equivalent to factoring the modulus, although the recent work of Boneh and Venkatesan [**19**] gives evidence that they may not be equivalent. There are a few public key cryptographic schemes for which one can prove that breaking the system is at least as hard as solving some hard mathematical problem, such as factoring a number or taking discrete logarithms. However, these are not discussed here.

We do note that for some classes of finite abelian groups one can prove that breaking the Diffie–Hellman key exchange protocol is polynomial time equivalent to solving a DLP. What is interesting about this work is that this result uses auxiliary groups which are themselves usually taken to be elliptic curves. The interested reader should consult [**94**], [**95**], [**18**] and Section IX.4.

The requirement in the signature schemes for a bijective function, f, from G to $\mathbb{Z}/(\#G)\mathbb{Z}$ may seem a little restrictive. For the groups, $\mathbb{F}_p^*$, the bijective function to use is obvious. For other groups the condition that f is bijective can be weakened. What is really required is a function

$$f : G \longrightarrow \mathbb{Z}/M\mathbb{Z}$$

for some number M, of the order of magnitude of the size of the group G, which is almost injective. In other words its degree as a map should be 'small'.

For elliptic curve systems the group elements are presented as pairs, (x, y), over some finite field. Such a pair represents a point on an elliptic curve. Over large prime fields, $\mathbb{F}_p$, field elements are naturally represented as integers modulo p, and one usually just uses the x-coordinate of the curve as the map from points (group elements) to integers modulo p (the latter prime turns out to be close to $\#G$, and is thus used for M above). This is a degree two map and will clearly suffice for applications. For large finite fields of characteristic two, one performs a similar method, but a way of converting the x-coordinate into an integer is needed. A simple method, used in practice, is to take the representation of x relative to a given basis of $\mathbb{F}_{2^n}$ over $\mathbb{F}_2$, and interpret the same coefficients as the binary digits of an integer. As long as Alice and Bob are using the same internal representation and order conventions, or at least Bob knows how to convert from his internal representation into Alice's, their implementations should be interoperable.

I.2. What Types of Group are Used

All of the above protocols work for a general abelian group, G, so one could consider various other groups to use in such protocols. However, since the protocols are to be implemented in hardware or software, the group operation should be simple to realize. One way of interpreting this condition, but not the only way, is to insist that the group operation be given by simple algebraic formulae. In other words G must be a commutative finite algebraic group. This then restricts quite considerably the types of such groups which are available.

A commutative finite algebraic group is essentially equivalent to the product of a finite number of copies of the additive and multiplicative groups of finite fields and a finite number of abelian varieties. For all practical purposes, the latter can be taken to be Jacobians of curves. It will be seen in Chapter V that, owing to a general purpose algorithm of Pohlig and Hellman, the group G should have a large subgroup of prime order. Thus we can restrict ourselves to only considering single copies of additive and multiplicative subgroups of finite fields or Jacobians.

The DLP in some additive groups is clearly easy, e.g. the additive group of a finite field. Fortunately, this is not the case, as far as is known, for the group of an elliptic curve. Not surprisingly, all of the above protocols were originally described in terms of the finite (multiplicative) abelian group $\mathbb{F}_q^*$. However, if one uses such groups the choice of q needs to be very large indeed, because there are known sub-exponential methods for solving the DLP in $\mathbb{F}_q^*$ (see [**1**] and [**88**]). These methods are usually based on the ideas behind the well known number field sieve factoring method (see [**77**]).

This situation led Miller [**103**] and Koblitz [**62**] to propose the technique, common in number theory, of replacing a group such as $\mathbb{F}_q^*$ with the group, $E(\mathbb{F}_q)$, of rational points on an elliptic curve, E, defined over $\mathbb{F}_q$ (these concepts will be precisely defined later). This technique will be seen again in the elliptic curve factoring method and the elliptic curve primality proving method. Elliptic curves are Jacobians of dimension one and so are the simplest case of a Jacobian. It turns out that the (additive) DLP in elliptic curve groups is, at present, orders of magnitude harder than the corresponding problem in the multiplicative group of a finite field of a similar size, a fact that is more precisely quantified in the next section.

If one wants to avoid algebraic groups then only one other type of group is known which is secure and almost practical. These are the class groups of orders of number fields. These were originally proposed by Buchmann and Williams [**23**] for class groups of imaginary quadratic orders. The protocols used in this situation differ slightly from those described earlier, but the essential features remain the same. In imaginary quadratic orders the elements of the class group can be represented by reduced binary quadratic forms. These forms can be multiplied using the standard composition and reduction algorithms which date back to Gauss (see [**29**] and [**50**]). We shall see in a later chapter that the arithmetic on an elliptic curve and in the Jacobian of a hyperelliptic curve is closely related to this composition of binary quadratic forms.

Such schemes based on class groups are particularly interesting, as breaking some of the proposed cryptosystems is provably as hard as factoring the discriminant of the order. However, the protocols are at present very slow owing to the complexity of the group operations. For other work on class group based systems, see [**10**], [**20**], [**22**] and [**52**].

There are cryptosystems based on elliptic curves which are provably as hard as known mathematical problems. For example there are systems based on elliptic curves over $\mathbb{Z}/n\mathbb{Z}$, where n is the product of two primes, for which the ability to break the system is as hard as factoring the modulus n (see the work of Meyer and Müller [**101**]). However, Joye and Quisquater [**57**] pointed out that the system of Meyer and Müller is reducible to the system of Rabin and Williams (see [**129**] and [**163**]). Hence, since the Meyer–Müller system is probably slower than the Rabin–Williams system, we shall not discuss the former system further.

There are other systems based on elliptic curves over $\mathbb{Z}/n\mathbb{Z}$, which are in some sense elliptic curve analogues of the RSA scheme (see for example Koyama et al. [**68**]). However, these are not provably as hard as factoring and they appear to offer no advantage over RSA in terms of security but do give a decrease in performance when compared with RSA. These schemes are not discussed further in this book. The reader is referred instead to [**17**], [**58**], [**70**], [**90**], [**121**] and [**159**].

I.3. What it Means in Practice

In this section we discuss the practical implications of using the group $E(\mathbb{F}_q)$ of a suitably chosen elliptic curve over a finite field to implement a DLP-based cryptosystem, as opposed to the more 'conventional' choice of the multiplicative group $\mathbb{F}_p^*$ of a finite field. Notice that, in the comparison, $\mathbb{F}_q$ and $\mathbb{F}_p$ need not be the same field. The key observation is that, for a well-chosen curve (in a sense to be made clear later in the book), the best known method for solving the DLP on $E(\mathbb{F}_q)$ is of complexity exponential in the size $n = \lceil \log_2 q \rceil$ of the field elements, while algorithms that are sub-exponential in $N = \lceil \log_2 p \rceil$ are available for the DLP in $\mathbb{F}_p^*$.

More specifically, the best known general algorithms for the elliptic curve DLP are of complexity proportional to

$$C_{\mathrm{EC}}(n) = 2^{n/2}$$

(see Chapter V).

Define the function

$$L_p(v, c) = \exp\left(c(\log p)^v (\log \log p)^{(1-v)}\right),$$

where 'log' without base specification denotes real natural logarithms. When $v = 1$, the function L_p is exponential in $\log p$, while for $v = 0$ it is polynomial in $\log p$. When $0 < v < 1$, the behaviour is strictly between polynomial and exponential, and is referred to as *sub-exponential*.

Discrete logarithms in $\mathbb{F}_p$ can be found in time proportional to $L_p(1/3, c_0)$, where $c_0 = (64/9)^{1/3} \approx 1.92$, using a general number field sieve method ([**99**,

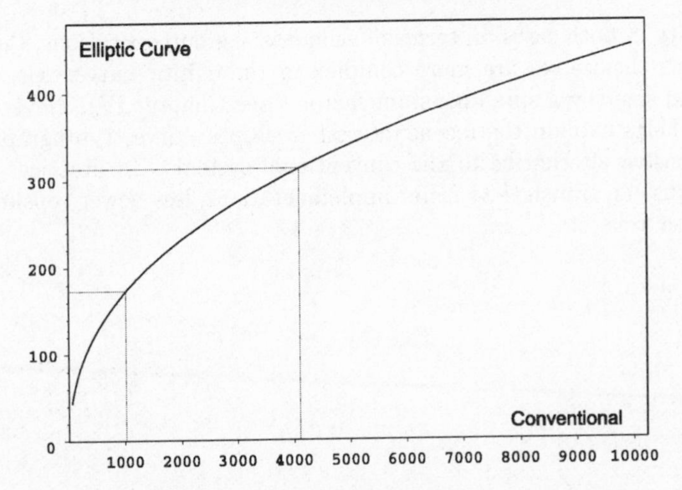

FIGURE I.1. Elliptic curve vs. conventional cryptosystem key sizes (in bits) for similar strength.

Ch. 3] [**114**]). In terms of N, and neglecting constant factors, the complexity is

$$C_{\text{CONV}}(N) = \exp(c_0 N^{1/3}(\log(N \log 2))^{2/3}),$$

where the subscript CONV stands for 'conventional'. Notice that the best known algorithms for *integer factorization* are of roughly the same asymptotic complexity (see [**99**] and [**77**]). Therefore, the discussion and comparisons in what follows apply also to conventional public key cryptosystems based on factorization, e.g., RSA.

Equating C_{EC} and C_{CONV} (and, again, neglecting constant factors in the complexities), it follows that for similar levels of security, we must have

$$n = \beta N^{1/3} (\log(N \log 2))^{2/3}$$

where $\beta = 2c_0/(\log 2)^{2/3} \approx 4.91$. Now, the parameters n and N can be interpreted as the 'key sizes', in bits, for the respective cryptosystems. Therefore, with current algorithmic knowledge, the key size in an elliptic curve cryptosystem grows slightly faster than the *cube root* of the corresponding 'conventional' key size, for similar cryptographic strength.

The relation is plotted in Figure I.1, where the correspondence for 'conventional' key sizes of 1024 and 4096 bits (common values for RSA) has been emphasized with the dotted lines. The equivalent key sizes shown for an elliptic curve cryptosystem are 173 and 313 bits, respectively. Given that various approximations have been used, and various constants neglected, such figures are, of course, approximate and give only general trends. A fair comparison should also take into account the complexity of implementing the cryptosystem. While the implementation of group exponentiation is of about the same complexity in both cases, in terms of elementary group operations, the group operations themselves are more complex in the elliptic curve case, for the same field size (by a small constant factor – see Chapter IV). Nevertheless, the plot helps explain the recent interest in elliptic curve cryptography as a less expensive alternative to the conventional systems. In practice, shorter key lengths can translate to faster implementations, less power consumption, less silicon area, etc.

CHAPTER II

Finite Field Arithmetic

When implementing an elliptic curve system an important consideration is how to implement the underlying field arithmetic. The problems encountered in such implementations are addressed in this chapter, with attention being focused on questions which arise mostly in software implementations, although some hardware issues are briefly mentioned. Two questions of particular importance are whether to use even or odd characteristic fields and secondly, whether to restrict implementation to fields of a special type, for efficiency, or support any type of finite field.

II.1. Fields of Odd Characteristic

In this section, implementation of arithmetic in $\mathbb{F}_p$, where p is a 'large' prime, will be considered. Field elements will be naturally represented as integers in the range $0, 1, \ldots, p-1$, with the usual arithmetic modulo p. General techniques for handling multi-precision integers are not discussed, as they are treated very effectively elsewhere, e.g. [**61**]. However, we will focus on techniques for performing fast modular arithmetic.

We need to implement the four standard arithmetic operations in $\mathbb{F}_p$, namely addition, subtraction, multiplication and division. It is, however, the last two of these (and particularly the last) which produce the most challenge. In what follows let w denote the *word size* of the basic word, in bits, and $b = 2^w$ the corresponding *base*. For example, typical values in present-day computers are $w = 32$ and $w = 64$. The integer b will be the base used in expressing multi-precision integers. However, some implementations for multi-precision integers make use of different bases. Two common alternatives are:

A power of ten. These are very inefficient since powers of ten are not the natural arithmetic for performing calculations within a computer. A power of ten is usually chosen to make input and output of the large integers easier. This, however, is poor programming practice as very little time spent by a program will be in the input and output phase. Most of the time will be spent in calculations, where a base which is a power of two will be much more efficient.

A base of half the word size. If we choose a base of $b = 2^{16}$, or $b = 2^{32}$, where b^2 is now the base corresponding to the full word size, then some advantages accrue. The basic multiplication step of the coefficients in the base b representation of the multi-precision integer can be performed very easily in a language like C. This is because the result of multiplying two base b integers will still fit in a word, with no code to cope with the carry being needed. But now, twice as many iterations need to be carried out for even a simple operation such as addition, and the situation is worse for operations where the algorithms used in practice are of non-linear complexity, such as multiplication.

The problem leading to the use of half-words can be alleviated by having a function, implemented in a small amount of machine code, which implements the operation of multiplying two full word size integers. Such a function would return the upper, u, and lower, l, portions in

$$ub + l = w_1 \times w_2,$$

where $w_1, w_2 < b$. For many processors this either is implemented on chip, for example on the Pentium®, or can be performed efficiently using the floating point coprocessor, available on some machines. Given that this is such a basic operation it is very important that it is implemented as efficiently as possible. The trouble of having to rewrite a few lines of machine code for every target architecture is a small price to pay for the large increase in speed which results.

II.1.1. Moduli of a special form/precomputed moduli. One possible choice of 'special moduli' are those of the form $p = b^t - a$, for some 'small' value of a. Such moduli are discussed in [**37**] and [**99**]. The following arithmetic is described in the latter reference. The algorithm for modular multiplication uses the standard multi-precision multiplication routines followed by a fast reduction procedure.

ALGORITHM II.1: **Reduction Modulo $p = b^t - a$.**

INPUT: An integer x.
OUTPUT: $r = x \pmod p$.
1. $q_0 \leftarrow \lfloor x/b^t \rfloor$, $r_0 \leftarrow x - q_0 b^t$, $r \leftarrow r_0$, $i \leftarrow 0$.
2. While $q_i > 0$ do:
3. $\quad q_{i+1} \leftarrow \lfloor q_i a/b^t \rfloor$, $r_{i+1} \leftarrow q_i a - q_{i+1} b^t$,
4. $\quad i \leftarrow i + 1$, $r \leftarrow r + r_i$.
5. While $r \geq p$ do $r \leftarrow r - p$.
6. Return r.

First note that the quotient on division of an n-word number by a power of b is easy to compute by shifting the numerator to the right a fixed number of words. Also note that subtraction of a multiple of a power of b is nothing but a subtraction of a number shifted to the left a given number of words. The reduction procedure is therefore performed using only shift and add operations and a multiplication by a. It therefore removes the need for any divisions to occur.

Modular inversion is often performed using the Euclidean algorithm and is therefore generally slow compared to a multiply. Fortunately, for elliptic curve cryptosystems, one can implement the underlying curve arithmetic to remove the need for almost all field inversions. We will elaborate on this issue in Chapter IV. Further improvements can be obtained if a is chosen to have low weight, by which we mean that the binary representation of a has only a few non-zero bits. This case is discussed, in the context of binary polynomial arithmetic, in Section II.2.1. Similar considerations apply to integers.

Another method, akin to using moduli of a special form, is to use precomputed tables for performing the modular reduction. Although the prime modulus need not have special form, it will probably be selected at startup of the system. The precomputation of the tables can then be performed once and for all. The disadvantage from a cryptographic viewpoint is that every person using the system will have to use curves defined over the same finite field. This raises problems of interoperability. For example if one bank is signing a document for passing to another bank then both banks will need to use the same field. This implies a level of agreement and standardization not yet generally available, although standards are currently being drafted to address these issues.

In addition, using fixed moduli of special form may not be preferred for security reasons. If everyone is forced to use the same field it becomes an attractive target for cryptanalysts. Breaking such a system, perhaps using special properties of the particular field which might be discovered, then has even more serious consequences than otherwise. Although this may give an attractive target for cryptanalysts one should not overestimate their capabilities. Nevertheless, prudent cryptographic practice would suggest changing the system parameters on some regular basis to discourage the attack of any particular parameter set.

II.1.2. Residue number system arithmetic. Residue number system (RNS) arithmetic is a very old idea which relies on the Chinese Remainder Theorem (CRT). Suppose we wish to work with a modulus p. A set of auxiliary primes, p_i (of word size), are chosen such that

$$\prod_{i=1}^{t} p_i > p^2.$$

We then represent an element, x, modulo p as the vector $(x_1, \ldots, x_t)$, where

$$x \equiv x_i \pmod{p_i}.$$

What makes this particularly appealing is that to add and multiply such numbers we need only compute the addition and multiplication of their components, of size very much smaller than the original modulus. The final result is obtained by the CRT.

As an example consider $p = 10727311963$ and $x = 1213212$, and assume we wish to work with 16-bit words. We take the primes $p_1 = 65521$, $p_2 = 65519$, $p_3 = 65497$, $p_4 = 65479$, and $p_5 = 65449$. We then represent x in this residue number system as

$$x \equiv (33834, 33870, 34266, 34590, 35130).$$

We can then compute $x + x$ and x^2 using simple word length arithmetic operations and find that

$$x + x \equiv (2147, 2221, 3035, 3701, 4811),$$

and

$$x^2 \equiv (22165, 4729, 59534, 35812, 10556).$$

However, we still need to perform the reduction operation for both addition and multiplication. This is particularly difficult using RNS arithmetic as it is hard to compare the size of elements and to perform integer division. Nevertheless, much recent work has been done in this area in the context of a sub-procedure for the number field sieve (NFS) algorithm (see [**32**]) and in the context of efficient hardware implementations for RSA-based systems (see [**126**] and [**127**]).

II.1.3. Barrett reduction. When using Barrett reduction, the standard multi-precision methods are used for integer arithmetic operations. However, the modular reduction is performed in a rather efficient way. We assume we are given a positive integer x which is of size at most p^2. We wish to compute $x \pmod p$. As a precomputation we compute

$$\mu = \lfloor b^{2t}/p \rfloor$$

where $b^t > p > b^{t-1}$ and b once again is the base size of the computer. We give the algorithm for computing $x \pmod p$ from [**99**] and leave the reader to consult that book for a justification.

ALGORITHM II.2: **Barrett Reduction.**

INPUT: x, p and μ such that $x < b^{2t}$, $b^{t-1} < p < b^t$ and $\mu = \lfloor b^{2t}/p \rfloor$.
OUTPUT: $z = x \pmod p$.

1. $q_0 \leftarrow \lfloor x/b^{k-1} \rfloor$.
2. $q \leftarrow \lfloor (\mu q_0)/b^{k+1} \rfloor$.
3. $r_1 \leftarrow x \pmod{b^{k+1}}$, $r_2 \leftarrow qp$.
4. $z \leftarrow r_1 - r_2$.
5. If $z < 0$ then $z \leftarrow z + b^{k+1}$.
6. While $z \geq p$ do $z \leftarrow z - p$.
7. Return z.

The only complicated part of this reduction is the computation of μq in Step 2. On the face of it this appears to be a full multi-precision multiplication.

However, on a second glance we see that the least significant words of this multiplication need not be computed (see [**99**, Ch. 14] for more details).

II.1.4. Montgomery arithmetic. By far the most successful way to implement arithmetic modulo a large prime p is to use a representation due to Montgomery [**105**]. Again assume b is the word base. Define t and R by

$$R = b^t > p.$$

Every element $x \in \mathbb{F}_p$ is represented by $xR \pmod{p}$. There is clearly a one-to-one relationship between this latter representation and the usual one. Addition and subtraction in this Montgomery representation can be performed in the usual way; however, multiplication is much faster. Our treatment will again follow that in the *Handbook of Applied Cryptography* [**99**, Ch. 14].

Before Montgomery multiplication is explained, the process of Montgomery reduction is considered. This is a procedure which takes as input an integer y with $0 \leq y < pR$ and returns $yR^{-1} \pmod{p}$.

LEMMA II.1. *In such a situation, if we set* $u = -yp^{-1} \pmod{R}$ *and* $x = (y + up)/R$ *then* x *is an integer such that* $x < 2p$ *and* $x \equiv yR^{-1} \pmod{p}$.

PROOF. Clearly the last modular equality holds since

$$x = (y + up)R^{-1} \equiv yR^{-1} \pmod{p}.$$

So we need to show that x is indeed an integer, but this is clear by definition of u. To see that $x < 2p$ notice that

$$x = (y + up)/R < (pR + Rp)/R = 2p.$$

□

A summary of the procedure is given in Algorithm II.3.

ALGORITHM II.3: **Montgomery Reduction (Simple Case).**

INPUT: A number $y < pR$.
OUTPUT: $x = yR^{-1} \pmod{p}$.
1. $u \leftarrow -yp^{-1} \pmod{R}$.
2. $x \leftarrow (y + up)/R$.
3. If $x \geq p$ then $x \leftarrow x - p$.
4. Return x.

To compute x we first compute u, which is easy once $p^{-1} \pmod{R}$ has been computed, since R is a power of the word base. Thus u can be computed with no modular divisions, the computation of $p^{-1} \pmod{R}$ being done once and stored for later use. To compute z we need to perform one further multi-precision multiplication, a multi-precision addition and then a division by R. But division by R is a simple shift of the words in $y + up$ to the right by t spaces, since $R = b^t$.

In fact we can be even more efficient by setting $p' = -p^{-1} \pmod{b}$ and if y is given by

$$y = (y_{2t-1}, \ldots, y_1, y_0)_b = y_{2t-1}b^{2t-1} + \cdots + y_1 b + y_0$$

then $yR^{-1} \pmod{p}$ can be computed by performing the following steps from [99, Ch. 14]:

ALGORITHM II.4: **Montgomery Reduction.**

INPUT: A number $y < pR$.
OUTPUT: $z = yR^{-1} \pmod{p}$.
1. For $i = 0$ to $t - 1$ do:
2. $u \leftarrow y_i p' \pmod{b}$,
3. $y \leftarrow y + upb^i$.
4. $z \leftarrow y/R$.
5. If $z \geq p$ then $z \leftarrow z - p$.
6. Return z.

The operation in Step 2 above can be implemented using a single precision multiplication operation. The operation in Step 3 is performed using a shift, a scalar multiplication and then an add. It has already been remarked that $z \leftarrow y/R$ is performed using a shift of y to the right by t words. Hence to compute $yR^{-1} \pmod{p}$ we need only perform a set of multi-precision addition, scalar multiplication and shift operations.

To compute $p' = -p^{-1} \pmod{b}$, the extended Euclidean algorithm can be used. However, it is rather easy to compute $x^{-1} \pmod{2^w}$ using the following algorithm.

ALGORITHM II.5: **Computing** $x^{-1} \pmod{2^w}$.

INPUT: An odd integer x, $0 < x < 2^w$.
OUTPUT: $y = x^{-1} \pmod{2^w}$.
1. $y \leftarrow 1$.
2. For $i = 2$ to w do:
3. If $2^{i-1} < xy \pmod{2^i}$ then $y \leftarrow y + 2^{i-1}$.
4. Return y.

To verify the correctness of Algorithm II.5, note that at the end of every execution of Step 3 we have

$$xy \equiv 1 \pmod{2^i}$$

(with the initial condition corresponding to $i = 1$). The method is very efficient, as only single precision arithmetic is used, assuming $2^w \leq b$ which holds in our case. This method of computing $x^{-1} \pmod{2^w}$ is due to Dussé and Kaliski [38].

Suppose two elements $x, y \in \mathbb{F}_p$ are given in their Montgomery representation, i.e. $X = xR \pmod{p}$ and $Y = yR \pmod{p}$. To compute $Z = xyR \pmod{p}$, first compute the standard multi-precision multiplication of X and Y to obtain $Z' = xyR^2$ which is a number of size at most $p^2 < pR$. By applying Montgomery reduction to the number Z' we obtain Z. Thus to multiply two elements in Montgomery representation we need only perform a single multi-precision multiplication followed by a Montgomery reduction. No divisions are needed.

The operation can be made more efficient by 'interleaving' the multiplication and reduction steps. Assume that X and Y are given in the form above, i.e. $(x_{t-1}, \dots, x_0)_b$ and $(y_{t-1}, \dots, y_0)_b$. To compute the Montgomery product $Z = XYR^{-1} \pmod{p}$, perform the following:

ALGORITHM II.6: **Montgomery Multiplication.**

INPUT: X and Y as above.
OUTPUT: $Z = XYR^{-1} \pmod{p}$.

1. $Z \leftarrow 0$.
2. For $i = 0$ to $t-1$ do:
3. $\quad u \leftarrow (z_0 + x_i y_0)p' \pmod{b}$,
4. $\quad Z \leftarrow (Z + x_i y + up)/b$.
5. If $Z \geq p$ then $Z \leftarrow Z - p$.
6. Return Z.

Notice that the computation of u in Step 3 can be performed in single precision and the computation of Z in Step 4 requires two multiplications of a multi-precision integer by a word, then two multi-precision additions followed by a right shift.

Division in Montgomery representation can be performed using the binary extended Euclidean algorithm (see, e.g., [**61**], [**29**]). For example, given $X = xR \pmod{p}$, we can compute, using the standard binary extended Euclidean algorithm, the number $Y = x^{-1}R^{-1} \pmod{p}$. Then to compute $x^{-1}R \pmod{p}$ we need only perform a Montgomery multiplication of Y and $R^3 \pmod{p}$.

Kaliski noticed [**58**] that the binary extended Euclidean algorithm can be modified to compute the Montgomery inverse. Kaliski defines the Montgomery inverse of a number x to be the integer $x^{-1}R \pmod{p}$, which is not quite what we want, but it is useful in some contexts.

II.1.5. Solving quadratic equations in fields of odd characteristic. Solving quadratic equations is an important operation in the context of elliptic curves, where it is used to obtain the y-coordinate of a point given its x-coordinate. In fields of characteristic different from two, this is done through the usual school formula, so the problem reduces to that of extracting square roots. The problem for the case of a prime finite field $\mathbb{F}_p$, $p > 2$, is considered.

Assume we wish to solve the equation

$$x^2 \equiv a \pmod{p}.$$

To test whether such an equation actually has a solution, the Legendre symbol $\left(\frac{a}{p}\right)$, which is equal to 1 if a is a square modulo p, 0 if $a \equiv 0 \pmod{p}$, or -1 otherwise, is used. To compute the Legendre symbol the following method, based on quadratic reciprocity, can be used.

ALGORITHM II.7: **Legendre Symbol.**

INPUT: a and p.
OUTPUT: $\left(\frac{a}{p}\right) \in \{1,0,-1\}$.

1. If $a \equiv 0 \pmod{p}$ then return 0.
2. $x \leftarrow a$, $y \leftarrow p$, $L \leftarrow 1$.
3. $x \leftarrow x \pmod{y}$.
4. If $x > y/2$ then do:
5. $\quad x \leftarrow y - x$,
6. $\quad$ If $y \equiv 3 \pmod{4}$ then $L \leftarrow -L$.
7. While $x \equiv 0 \pmod{4}$ do $x \leftarrow x/4$.
8. If $x \equiv 0 \pmod{2}$ then do:
9. $\quad x \leftarrow x/2$,
10. $\quad$ If $y \equiv \pm 3 \pmod{8}$ then $L \leftarrow -L$.
11. If $x = 1$ then return L.
12. If $x \equiv 3 \pmod{4}$ and $y \equiv 3 \pmod{4}$ then $L \leftarrow -L$.
13. Swap x and y and go to Step 3.

Alternatively we could compute $a^{(p-1)/2} \pmod{p}$. It can thus be decided whether a is or is not a square. If $a \equiv 0 \pmod{p}$ then a has only one square root modulo p, which is 0. If $\left(\frac{a}{p}\right) = 1$ then there are two square roots modulo p and we need to determine one of them. The following algorithm is based on a method of Tonelli and Shanks(see **[29]**).

ALGORITHM II.8: **Square Root Modulo p.**

INPUT: a and p such that $\left(\frac{a}{p}\right) = 1$.
OUTPUT: x such that $x^2 \equiv a \pmod{p}$.

1. Choose random n until one is found such that $\left(\frac{n}{p}\right) = -1$.
2. Let e, q be integers such that q is odd and $p - 1 = 2^e q$.
3. $y \leftarrow n^q \pmod{p}$, $r \leftarrow e$, $x \leftarrow a^{(q-1)/2} \pmod{p}$,
4. $b \leftarrow ax^2 \pmod{p}$, $x \leftarrow ax \pmod{p}$.
5. While $b \not\equiv 1 \pmod{p}$ do:
6. $\quad$ Find the smallest m such that $b^{2^m} \equiv 1 \pmod{p}$,
7. $\quad t \leftarrow y^{2^{r-m-1}} \pmod{p}$, $y \leftarrow t^2 \pmod{p}$, $r \leftarrow m$,

```
8.       x ← xt (mod p), b ← by (mod p).
9. Return x.
```

An analogue of the above method can be used to take square roots in any group of even order.

II.2. Fields of Characteristic Two

Finite fields of characteristic 2 are attractive to implementers due to their 'carry-free' arithmetic, and the availability of different equivalent representations of the field, which can be adapted and optimized for the computational environment at hand.

Specifically, in this section we discuss arithmetic over the finite field $\mathbb{F}_{2^n}$, $n \geq 1$. Field elements are represented as binary vectors of dimension n, relative to a given basis $(\alpha_0, \alpha_1, \dots, \alpha_{n-1})$ of $\mathbb{F}_{2^n}$ as a linear space over $\mathbb{F}_2$. Field addition and subtraction are implemented as component-wise exclusive OR (XOR), while the implementations of multiplication and inversion depend on the basis chosen. Common practical choices and their implementations are discussed in the following sections. Polynomial, normal and subfield bases, plus some variants on these, are considered.

II.2.1. Polynomial bases. A *polynomial* (or *standard*) basis is of the form $(1, \alpha, \alpha^2, \dots, \alpha^{n-1})$, where α is a root of an irreducible polynomial $f(x)$ of degree n over $\mathbb{F}_2$. The field is then realized as $\mathbb{F}_2[x]/\langle f(x)\rangle$, and the arithmetic is that of polynomials of degree at most $n-1$, modulo $f(x)$.

Modular reduction. By choosing $f(x)$ as a *low weight* polynomial, i.e. one with the least possible number of non-zero coefficients, reduction modulo $f(x)$ becomes a very simple operation that is performed in time $O(Wn)$, where W is the weight of f. It turns out that for cases of practical interest, it can be assumed that $f(x)$ is either a trinomial or a pentanomial (i.e., $W = 3$ or 5). The existence, distribution and other properties of irreducible trinomials over $\mathbb{F}_2$ have been extensively studied in the literature. In particular, it follows from a theorem of Swan [**156**] that irreducible trinomials do not exist for $n \equiv 0 \pmod 8$, and that they are rather scarce when $n \equiv 3$ or $5 \pmod 8$ – see also [**9**, Ch. 6], [**47**], [**86**, Ch. 3], and the many references therein. Empirical studies for values of n into the thousands ([**14**] [**144**]) show that irreducible trinomials exist for over half of the values of n covered. On the other hand, the table in [**144**] shows that, at least up to degree n=10 000, in all cases where an irreducible trinomial is not available, an irreducible pentanomial is. In fact, there is no known value of n for which an irreducible polynomial of odd weight $W \leq 5$ does not exist. The general question, however, remains open for any fixed odd weight $W > 3$.

The following algorithm exemplifies reduction of a polynomial of degree $2n-2$, such as is obtained from the product of two polynomials of degree $n-1$, modulo a trinomial $f(x)$. The extension to pentanomials is straightforward.

ALGORITHM II.9: **Reduction Modulo** $f(x) = x^n + x^t + 1, \ 0 < t < n.$

```
INPUT:  a(x) = a_0 + a_1 x + a_2 x^2 + ··· + a_{2n-2} x^{2n-2} ∈ F_2[x].
OUTPUT: r(x) ≡ a(x) (mod f(x)),  deg r(x) < n.
1.  For i = 2n-2 to n by -1 do:
2.        a_{i-n} ← a_{i-n} + a_i,  a_{i-n+t} ← a_{i-n+t} + a_i.
3.  Return r(x) = a_0 + a_1 x + a_2 x^2 + ··· + a_{n-1} x^{n-1}.
```

The above algorithm operates on $a(x)$ 'in place', obviating the need for extra storage for the result $r(x)$. Also, in a software environment, the algorithm is easily adapted to operate on computer words. If $n-t \geq w$, where w is the word size, then the algorithm scans the words containing the coefficients $a_{2n-2}, a_{2n-1}, \ldots, a_n$, from higher order to lower order, adding each word into two positions offset $n-t$ and n bits back, respectively. The condition on $n-t$ guarantees that a word does not add to any part of itself, and is thus processed only once. Each offset location requires up to two word XOR operations, since it might not be word-aligned. The total number of word XOR operations in the trinomial case is therefore at most $4\lceil n/w \rceil$. In general, reduction modulo an irreducible of weight W requires at most $2(W-1)\lceil n/w \rceil$ word XOR operations.

Another favoured choice of irreducible polynomial is one of the form $f(x) = x^n + g(x)$ where the degree of $g(x)$ is 'small' relative to n. This is analogous to the choice of primes of the form $p = b^t - a$ for $\mathbb{F}_p$ in Section II.1.1, for a small value of a. This case also leads to a fast modular reduction procedure, although slightly less efficient than the one based on low weight irreducibles.

Multiplication. When using polynomial bases, the first stage in computing the product of two elements of $\mathbb{F}_{2^n}$ is the multiplication of two polynomials of degree at most $n-1$ in $\mathbb{F}_2[x]$. This is a 'carry-free' version of the multiplication of two n-bit integers, and most methods for large integer multiplication have analogues in $\mathbb{F}_2[x]$. In particular, a fast asymptotic method of complexity $O(n \log n \log\log n)$, due to Schönhage, is described in [**140**] (see also [**61**, Ch. 4]). However, in practical implementations of elliptic curve cryptography, moderate values of n in the low hundreds are typical. When data is appropriately packed into computer words (typically, of 32 or 64 bits), this translates into a small number of words, and the overhead of the fast asymptotic methods is seldom justified. Instead, simpler methods are often used, which are asymptotically inferior but lend themselves to very compact and efficient implementations for the range of values of interest. In particular, the old and well tried $O(n^{\log_2 3})$ *recursive subdivision* method first described for integers by Karatsuba [**59**] is often appropriate.

Assume n is even. To compute the product $a(x)b(x)$, where $a(x), b(x) \in \mathbb{F}_2[x]$ have degree $n-1$, we write

$$a(x)b(x) = (A_1(x)X + A_0(x))(B_1(x)X + B_0(x)), \qquad \text{(II.1)}$$

where A_0, A_1, B_0, B_1 are polynomials of degree $n/2-1$, and $X = x^{n/2}$. The right-hand side of Equation (II.1) can be regarded as the product of two linear polynomials in X, with coefficients in $\mathbb{F}_2[x]$. This product can be derived from the three products A_0B_0, A_1B_1 and $(A_0 + A_1)(B_0 + B_1)$ of polynomials of degree $n/2-1$; i.e., one problem of size n is solved by solving three problems of size $n/2$. Similarly, when n is odd, a problem of size n can be reduced to one of size $(n-1)/2$ and two of size $(n+1)/2$. In either case, proceeding recursively leads to an overall number of operations $O(n^{\log_2 3})$ (detailed analysis can be found in [**61**, Ch. 4]). In practice, the procedure is implemented on words, and the multiplication of two word-sized polynomials, taken as the basis of the recursion, is optimized for the machine at hand. If n is fixed, the recursion can be 'unrolled', and the computation can be expressed as a straight-line algorithm. Also, it is sometimes advantageous to depart from a pure binary subdivision recursion. For example, multiplying two three-word polynomials takes seven word multiplications using a pure Karatsuba procedure, but can be done in six word multiplications using a direct straight-line algorithm for multiplication of binary quadratic polynomials – in fact, six is the minimum number of multiplications for this problem (see, e.g., [**164**], [**76**]).

As a final remark on multiplication in polynomial representation, recall that squaring is much easier than general multiplication in $\mathbb{F}_2[x]$. To square a polynomial, we just 'thin it out', inserting a zero between every two original binary coefficients. Thus, the complexity of the squaring operation is $O(n)$, comparable to that of the modular reduction, assuming a low weight modulus is used.

Inversion. The extended Euclidean algorithm is a natural choice for computing inverses in polynomial representations. As with multiplication, fast asymptotic algorithms are available for this computation. An $O(M(n)\log n)$ algorithm for computing greatest common divisors is described in [**5**], where $M(n)$ denotes the complexity of multiplying n-bit polynomials. The algorithm can be easily adapted to compute modular inverses and, combined with the results of [**140**], yields an overall complexity $O(n \log^2 n \log\log n)$. But again, asymptotically fast methods start being effective at fairly large values of n, usually beyond those used in practical elliptic curve cryptosystems. Therefore, practical implementations for moderate values of n often rely on variants of the binary extended Euclidean algorithm for polynomials (see, e.g., [**9**, Ch. 2], [**61**, Ch. 4]).

In any case, $\mathbb{F}_{2^n}$ inversion is often significantly slower than multiplication. In fact, an inversion can sometimes be favourably replaced by a chain of multiplications. Such schemes derive from the field equation, which can be recast as

$$\beta^{-1} = \beta^{2^n-2} = \left(\beta^{2^{n-1}-1}\right)^2,$$

for all $\beta \neq 0$ in $\mathbb{F}_{2^n}$. A technique for minimizing the number of general multiplications in this computation (i.e., not counting squarings, which are much cheaper) is described by Itoh and Tsujii in [**54**]. The method is based on the identities

$$\beta^{2^{n-1}-1} = \begin{cases} \beta^{(2^{\frac{n-1}{2}}-1)(2^{\frac{n-1}{2}}+1)} = \left(\beta^{2^{\frac{n-1}{2}}-1}\right)^{2^{\frac{n-1}{2}}} \beta^{2^{\frac{n-1}{2}}-1}, & n \text{ odd}, \\ \beta\beta^{2^{n-1}-2} = \beta\left(\beta^{2^{n-2}-1}\right)^2, & n \text{ even}. \end{cases}$$

Denoting by $\mu(n-1)$ the number of multiplications required to compute β^h where $h = 2^{n-1} - 1$, we have $\mu(n-1) = 1 + \mu((n-1)/2)$ when n is odd, and $\mu(n-1) = 1 + \mu(n-2) = 2 + \mu((n-2)/2)$ when n is even. Now, setting $\mu(1) = 0$, $\mu(2) = 1$ as the basis for the recursion, it is straightforward to prove that $\mu(n-1) = \lfloor \log_2(n-1) \rfloor + W(n-1) - 1$, where $W(k)$ denotes the weight (number of non-zero bits in the binary representation) of a positive integer k. Also, the number of squarings is readily determined, using a simple recursion, to be $n-1$. As an example, consider $n = 163$. Then, since $162 = (10100010)_2$, we have $\mu(162) = 7+3-1 = 9$, i.e. an inverse in $\mathbb{F}_{2^{163}}$ can be computed with 9 multiplications and 162 squarings.

Clearly, the inversion scheme just described is advantageous when squaring is very inexpensive, as in the case when normal bases are used. (These bases are discussed in the next subsection.) The scheme might still be appropriate for polynomial bases, but this is more dependent on specific implementation details. An alternative way to trade inversions for multiplications in the context of elliptic curve computations, without increasing the number of squarings so significantly, is to use *projective coordinates* for the elliptic curve points. With this approach, field inversions are deferred, and usually only one such operation is required at the end of a long sequence of multiplications. We will get back to projective coordinates in Chapter IV.

An analogue of Montgomery multiplication for fields of characteristic two is described in [**67**]. We shall not consider this technique here, as modular reduction is not a computational bottleneck in characteristic two when low weight irreducible polynomials are used.

II.2.2. Normal bases. A *normal basis* of $\mathbb{F}_{2^n}$ over $\mathbb{F}_2$ has the form $(\alpha, \alpha^2, \alpha^{2^2}, \ldots, \alpha^{2^{n-1}})$ for some $\alpha \in \mathbb{F}_{2^n}$. It is well known (see, e.g., [**86**, Ch. 2]) that such bases exist for all $n \geq 1$. Normal bases are useful mostly in hardware implementations. First, the field squaring operation is trivial in normal basis representation, as it amounts to just a cyclic shift of the binary vector representing the input operand. More importantly, normal bases allow for the design of efficient *bit-serial multipliers*, such as the one described by Massey and Omura in [**115**].

A measure of the hardware complexity of such a multiplier is given by the number C_α of ones in the $n \times n$ binary matrix $T = (T_{ij})$ defined by

$$\alpha^{1+2^i} = \sum_{j=0}^{n-1} T_{ij}\alpha^{2^j}, \quad 0 \le i \le n-1. \tag{II.2}$$

The matrix T completely determines the structure of multiplication for the normal basis, as it captures all the information on products of basis elements. It is clear that $C_\alpha \le n^2$. On the other hand, C_α satisfies the lower bound $C_\alpha \ge 2n-1$ [**112**]. When the lower bound is attained, α is said to generate an *optimal normal basis* (ONB). An alternative characterization states that α generates an ONB if and only if for all i_1, i_2, $0 \le i_1 < i_2 \le n-1$, there exist integers j_1, j_2 such that $\alpha^{2^{i_1}+2^{i_2}} = \alpha^{2^{j_1}} + \alpha^{2^{j_2}}$. It is easy to verify the definitions are equivalent.

The existence of optimal normal bases has been completely characterized in [**112**] and [**45**] (see also [**15**, Ch. 5]). In particular, an ONB of $\mathbb{F}_{2^n}$ over $\mathbb{F}_2$ exists if and only if one of the following conditions holds:

(i) $n+1$ is prime, and 2 is primitive in $\mathbb{F}_{n+1}$; then the n non-trivial $(n+1)$st roots of unity form an ONB of $\mathbb{F}_{2^n}$ over $\mathbb{F}_2$, called a *Type I* ONB;

(ii) $2n+1$ is prime, and *either*
 (1) 2 is primitive in $\mathbb{F}_{2n+1}$ *or*
 (2) $2n+1 \equiv 3 \pmod 4$ and the multiplicative order of 2 in $\mathbb{F}_{2n+1}$ is n; that is 2 generates the quadratic residues in $\mathbb{F}_{2n+1}$;

 then, $\alpha = \gamma + \gamma^{-1}$ generates an ONB of $\mathbb{F}_{2^n}$ over $\mathbb{F}_2$, where γ is a primitive $(2n+1)$st root of unity; this is called a *Type II* ONB.

Apart from their hardware complexity advantages, we mention a few of the other interesting properties of ONBs, which follow readily from the characterization above. First, an ONB is always *self-dual* [**86**, Ch. 3], i.e., $\mathrm{Tr}_{q|2}(\alpha_i\alpha_j) = 1$ if and only if $i = j$, where α_i, α_j denote arbitrary basis elements, and $\mathrm{Tr}_{q|2}(z)$ denotes the trace of $z \in \mathbb{F}_q$ over $\mathbb{F}_2$, $q = 2^n$. Second, when an ONB of $\mathbb{F}_{2^n}$ exists, it is unique. Finally, the matrix T can be constructed directly from properties of the residue classes of integers modulo $n+1$ (Type I) or $2n+1$ (Type II) [**15**, Ch. 5]. Thus, the field algebra can be realized directly without first requiring the construction of a binary irreducible polynomial of degree n.

The bit-serial multipliers that are very effective for ONBs in hardware do not always map nicely to efficient software implementations, as single bit operations are expensive in the latter. Also, while efficient bit-serial implementations are available for multiplication in ONB representation, they do not carry to inversion operations. It turns out, however, that by applying simple permutations, operations on ONB representations of both Types I and II can be handled through polynomial arithmetic, in a manner similar to the case of standard bases.

For Type I ONBs, we observe that the minimal polynomial of α is $f(x) = x^n + x^{n-1} + \cdots + x + 1$, and the set $\{\alpha, \alpha^2, \alpha^{2^2}, \ldots, \alpha^{2^{n-1}}\}$ is the same as the set $\{\alpha, \alpha^2, \alpha^3, \ldots, \alpha^n\}$. Therefore, for an element with coordinates $(a_0, a_1, \ldots, a_{n-1})$ in ONB representation, we can write

$$\sum_{i=0}^{n-1} a_i \alpha^{2^i} = \sum_{j=1}^{n} a_{\pi(j)} \alpha^j ,$$

where the bijection $\pi : \{1, 2, \ldots, n\} \to \{0, 1, \ldots, n-1\}$ is defined so that $\pi(j) = i$ whenever $2^i \equiv j \pmod{n+1}$. Thus, after suitable permutation, we can operate on elements in ONB representation as polynomials modulo $f(x)$, or even simpler, modulo $(x+1)f(x) = x^{n+1} + 1$. The latter will give results expressed in terms of $1, \alpha, \alpha^2, \ldots, \alpha^n$, which are brought back to the desired basis set by using, when needed, the equality $1 = \alpha + \alpha^2 + \cdots + \alpha^n$.

A similar, albeit slightly more involved transformation for Type II ONBs is described by Blake et al. in [**16**]. Write $p = 2n + 1$, where n satisfies Condition (ii) above, and let γ be a pth root of unity. Let Φ denote the vector space of all polynomials over $\mathbb{F}_2$ of the form $a(x) = \sum_{j=1}^{2n} a_j x^j$, where $a_j = a_{p-j}$ for $j = 1, 2, \ldots, n$. We call the elements of Φ *palindromic polynomials.* In a *palindromic representation* of $\mathbb{F}_{2^n}$, each field element corresponds to a palindromic polynomial. Addition is defined in the usual way, and the product of two palindromic polynomials $a(x), b(x) \in \Phi$ is the unique polynomial $c(x) \in \Phi$ such that

$$c(x) \equiv a(x) \cdot b(x) \pmod{x^p - 1} . \tag{II.3}$$

When we substitute $x = \gamma$ in $a(x)$, we obtain

$$a(\gamma) = \sum_{j=1}^{2n} a_j \gamma^j = \sum_{j=1}^{n} a_j (\gamma^j + \gamma^{-j}) .$$

It follows from Condition (ii) that for every $j \in \{1, 2, \ldots, n\}$, exactly one element in the pair $\{j, p-j\}$ can be written as 2^i modulo p, for some $0 \le i \le n-1$. Hence, we can write

$$a(\gamma) = \sum_{i=0}^{n-1} a_{2^i} (\gamma^{2^i} + \gamma^{-2^i}) = \sum_{i=0}^{n-1} a_{2^i} (\gamma + \gamma^{-1})^{2^i} = \sum_{i=0}^{n-1} a_{2^i} \alpha^{2^i}, \tag{II.4}$$

where all indices are taken modulo p. Equation (II.4) implies that, up to permutation, the elements $a_1, a_2, \ldots, a_n$ are the coefficients of the ONB representation of $a(\gamma)$. It follows from this simple relationship between the coefficients of $a(x)$ and the ONB representation of $a(\gamma)$ that arithmetic operations in ONB representation can be realized as polynomial operations modulo $x^p - 1$. In particular, inverses in ONB representation can be computed using the Euclidean algorithm.

As an example of the transformation for Type II ONBs, consider the case $n = 5$. It is readily verified that this case satisfies Condition (ii), with 2 being primitive modulo $p = 11$. We have $(2^0, 2^1, 2^2, 2^3, 2^4) \equiv (1, 2, 4, -3, 5) \pmod{11}$.

Thus, an element $a_0\alpha+a_1\alpha^2+a_2\alpha^4+a_3\alpha^8+a_4\alpha^{16}$ in ONB representation corresponds to the palindromic polynomial

$$a_0x+a_1x^2+a_3x^3+a_2x^4+a_4x^5+a_4x^6+a_2x^7+a_3x^8+a_1x^9+a_0x^{10}.$$

II.2.3. Subfield bases. When $n = n_1n_2$, we can regard $\mathbb{F}_{2^n}$ as an extension of degree n_2 of $\mathbb{F}_{2^{n_1}}$, and represent elements of $\mathbb{F}_{2^n}$ using a basis of the form $\{\alpha_i\beta_j : 0 \le i < n_1,\ 0 \le j < n_2\}$, where $\beta_0, \beta_1, \dots, \beta_{n_2-1}$ form a basis of $\mathbb{F}_{2^n}$ over $\mathbb{F}_{2^{n_1}}$, and $\alpha_0, \alpha_1, \dots, \alpha_{n_1-1}$ form a basis of $\mathbb{F}_{2^{n_1}}$ over $\mathbb{F}_2$. Thus, arithmetic can be done in two stages, with an 'outer' section doing operations on elements of $\mathbb{F}_{2^n}$ as vectors of symbols from $\mathbb{F}_{2^{n_1}}$; and an 'inner' section performing the operations on the symbols as binary words. Any combination of bases can be used, e.g., normal basis for the outer section, and polynomial basis for the inner one.

The subfield representation is particularly advantageous when n_1 is large enough so that n_2 is small, but n_1 is still small enough so that symbol operations can be made very fast in the computational environment at hand, e.g. by implementing the $\mathbb{F}_{2^{n_1}}$ arithmetic through look-up tables. Values of n_1 between, say, 4 and 16 are typical. The $\mathbb{F}_{2^n}$ inversion operation benefits the most from this structure, as the Euclidean algorithm is performed on much shorter polynomials, and the scheme benefits from the parallelization resulting from operations on symbols. Thus, typically, the gap between the running times of inversion and multiplication is smaller when a subfield representation is used, as compared to a polynomial basis over $\mathbb{F}_2$. The latter, of course, is a special case of subfield representation with $n_1 = 1$.

Inversion methods based on repeated multiplication can also be made more efficient when a subfield is available. Here, for any non-zero $\beta \in \mathbb{F}_{2^n}$, we can write

$$\beta^{-1} = \frac{\beta^{s-1}}{\beta^s}, \qquad \text{(II.5)}$$

where $s = (2^n - 1)/(2^{n_1}-1)$. The key observation is that β^s is in the subfield $\mathbb{F}_{2^{n_1}}$ (being the *norm* [**86**, Ch. 3] of β over $\mathbb{F}_{2^{n_1}}$). Hence, to compute β^{-1}, we obtain first β^{s-1} with an optimized addition chain (discussed in Chapter IV), and then β^s with an additional multiplication. The quotient in Equation (II.5) is finally obtained with an inverse in $\mathbb{F}_{2^{n_1}}$ and a scalar multiplication by the resulting subfield element. A scheme along these lines is analysed in [**49**].

Besides their advantages in implementing finite field arithmetic, subfields can help in two other central problems in elliptic curve cryptosystems: curves whose coefficients are in subfields allow for easier determination of the group order (as discussed in Section VI.4), and they offer 'shortcuts' for the important point multiplication operation (as discussed in Section IV.3). Unfortunately, behind the same nice algebraic structure that leads to these advantages could also lurk as yet undiscovered cryptographic weaknesses, as suspected by some researchers.

II.2.4. Solving quadratic equations in $\mathbb{F}_{2^n}$. An equation of the form $x^2 + \beta = 0$ is trivially solved in $\mathbb{F}_{2^n}$ by writing its (double) root x_0 explicitly as $x_0 = \beta^{2^{n-1}}$. Other non-trivial quadratic equations can always be brought to the canonical form

$$x^2 + x + \beta = 0. \tag{II.6}$$

This equation has solutions in $\mathbb{F}_{2^n}$ if and only if $\mathrm{Tr}_{q|2}(\beta) = 0$. If x_0 is such a solution, then so is x_0+1.

The procedure for finding a solution varies according to the parity of n. If n is odd, an explicit solution is given by $x_0 = \tau(\beta)$, where τ, the *half-trace* function, is defined by

$$\tau(\beta) = \sum_{j=0}^{(n-1)/2} \beta^{2^{2j}}. \tag{II.7}$$

It can be verified by direct inspection that $\tau(\beta)^2 + \tau(\beta) = \beta + \mathrm{Tr}_{q|2}(\beta)$, which verifies the solution when $\mathrm{Tr}_{q|2}(\beta) = 0$.

When n is even, the half-trace will not do, and a solution is found using the following procedure. Let $\delta \in \mathbb{F}_{2^n}$ be such that $\mathrm{Tr}_{q|2}(\delta) = 1$. Such an element can be obtained either by randomly drawing field elements until one of the right trace is found (with a probability of one half in each try), or by deterministically computing the traces of the basis elements $\alpha_0, \alpha_1, \ldots, \alpha_{n-1}$. At least one basis element must have trace one. In practice, computing these basis traces can be a good investment, as the vector

$$\mathbf{t} = (\mathrm{Tr}_{q|2}(\alpha_0), \mathrm{Tr}_{q|2}(\alpha_1), \ldots, \mathrm{Tr}_{q|2}(\alpha_{n-1}))$$

is useful for computing traces of arbitrary field elements. With δ at hand, a solution x_0 to Equation (II.6) is given by

$$x_0 = \sum_{i=0}^{n-2} \left(\sum_{j=i+1}^{n-1} \delta^{2^j} \right) \beta^{2^i}. \tag{II.8}$$

To verify that x_0 is indeed a solution, we compute

$$\begin{aligned} x_0^2 + x_0 &= \sum_{i=1}^{n-1} \left(\sum_{j=i+1}^{n} \delta^{2^j} \right) \beta^{2^i} + \sum_{i=0}^{n-2} \left(\sum_{j=i+1}^{n-1} \delta^{2^j} \right) \beta^{2^i} \\ &= \delta(\beta^{2^{n-1}} + \beta^{2^{n-2}} + \cdots + \beta^2) + (\delta^{2^{n-1}} + \delta^{2^{n-2}} + \cdots + \delta^2)\beta \\ &= \delta \mathrm{Tr}_{q|2}(\beta) + \beta, \end{aligned}$$

where the last equality follows from $\delta^{2^{n-1}} + \delta^{2^{n-2}} + \cdots + \delta^2 = \mathrm{Tr}_{q|2}(\delta) + \delta = 1 + \delta$. Thus, $x_0^2 + x_0 = \beta$ if and only if $\mathrm{Tr}_{q|2}(\beta) = 0$, as desired.

Virtually all the algorithms discussed in this chapter find application in the implementation of elliptic curve systems discussed in the remainder of the book.

CHAPTER III

Arithmetic on an Elliptic Curve

There is an extensive literature on elliptic curves. They arise naturally in many branches of mathematics and are closely linked with the theory of elliptic functions, from which they derive their name. In the recent past they have, for instance, been studied for theoretical uses in the solution to Fermat's Last Theorem, [**162**]. One notices immediately on studying elliptic curves that they are not ellipses, and hence a brief account of how the name arises is appropriate at this point.

Just as the arc lengths on a circle give rise to the trigonometric functions, sin, cos and tan, a similar study for ellipses leads one to consider *elliptic integrals.* These are integrals of the form

$$\int \frac{dx}{\sqrt{4x^3 - g_2x - g_3}}.$$

Such integrals are multi-valued on the complex numbers and are only well defined modulo a period lattice. One can hence consider the values taken by an elliptic integral to be on a torus. The 'inverse' function of an elliptic integral is a doubly periodic function called an *elliptic function.* Indeed all meromorphic doubly periodic functions on the complex numbers arise in this way.

It turns out that every doubly periodic function $\wp$ with periods that are independent over $\mathbb{R}$ satisfies an equation of the form

$$\wp'^2 = 4\wp^3 - g_2\wp - g_3, \tag{III.1}$$

for some constants g_2 and g_3. For future reference, such a function $\wp$ will be referred to as a Weierstrass $\wp$ function. If we consider the pair $(\wp, \wp')$ as being a point in space, then the solutions to Equation (III.1) provide a mapping from a torus (as $\wp$ is doubly periodic) to the curve

$$Y^2 = 4X^3 - g_2X - g_3.$$

This is an example of an elliptic curve (the 4 in front of the X^3 term is traditional in analytic circles – it can clearly be scaled away).

In this chapter we present the basic concepts from the theory of elliptic curves that are required for developing the material in the rest of the book. The treatment is far from comprehensive, of course.[1] The reader is referred

[1] 'It is possible to write endlessly on elliptic curves.' S. Lang, in the foreword to [**72**].

to [**147**] and [**148**] for a more comprehensive treatment, including most proofs missing here.

III.1. General Elliptic Curves

Let K be a field, $\overline{K}$ its algebraic closure, and K^* its multiplicative group. An elliptic curve over K will be defined as the set of solutions in the projective plane $\mathbb{P}^2(\overline{K})$ of a homogeneous *Weierstrass equation* of the form

$$E : Y^2Z + a_1XYZ + a_3YZ^2 = X^3 + a_2X^2Z + a_4XZ^2 + a_6Z^3, \quad \text{(III.2)}$$

with $a_1, a_2, a_3, a_4, a_6 \in K$. This equation is also referred to as the *long Weierstrass form.* Such a curve should be non-singular in the sense that, if the equation is written in the form $F(X,Y,Z) = 0$, then the partial derivatives of the curve equation $\partial F/\partial X$, $\partial F/\partial Y$, and $\partial F/\partial Z$ should not vanish simultaneously at any point on the curve.

Let $\hat{K}$ be a field satisfying $K \subseteq \hat{K} \subseteq \overline{K}$. A point (X,Y,Z) on the curve is $\hat{K}$-*rational* if $(X,Y,Z) = \alpha(\hat{X},\hat{Y},\hat{Z})$ for some $\alpha \in \overline{K}$, $(\hat{X},\hat{Y},\hat{Z}) \in \hat{K}^3 \setminus \{(0,0,0)\}$, i.e., up to projective equivalence, the coordinates of the point are in $\hat{K}$. The set of $\hat{K}$-rational points on E is denoted by $E(\hat{K})$. When the field of definition of the curve, K, is clear from the context, we will refer to K-rational points simply as *rational points.* The curve has exactly one rational point with coordinate Z equal to zero, namely $(0,1,0)$. This is the *point at infinity,* which will be denoted by $\mathcal{O}$.

For convenience, we will most often use the *affine* version of the Weierstrass equation, given by

$$E : Y^2 + a_1XY + a_3Y = X^3 + a_2X^2 + a_4X + a_6, \quad \text{(III.3)}$$

where $a_i \in K$. The $\hat{K}$-rational points in the affine case are the solutions to E in $\hat{K}^2$, and the point at infinity $\mathcal{O}$. For curves over the reals, this point can be thought of as lying infinitely far up the y-axis. We will switch freely between the projective and affine presentations of the curve, denoting the equation in both cases by E. For $Z \neq 0$, a projective point (X,Y,Z) satisfying Equation (III.2) corresponds to the affine point $(X/Z, Y/Z)$ satisfying Equation (III.3). In Chapter IV, we will consider a different projective representation which is convenient from a computational point of view.

Given an elliptic curve defined by Equation (III.3), it is useful to define the following constants for use in later formulae:

$$\left.\begin{aligned} &b_2 = a_1^2 + 4a_2, \quad b_4 = a_1a_3 + 2a_4, \quad b_6 = a_3^2 + 4a_6, \\ &b_8 = a_1^2a_6 + 4a_2a_6 - a_1a_3a_4 + a_2a_3^2 - a_4^2, \\ &c_4 = b_2^2 - 24b_4, \quad c_6 = -b_2^3 + 36b_2b_4 - 216b_6. \end{aligned}\right\} \quad \text{(III.4)}$$

The *discriminant* of the curve is defined as

$$\Delta = -b_2^2b_8 - 8b_4^3 - 27b_6^2 + 9b_2b_4b_6.$$

When $\mathrm{char}(K) \neq 2, 3$ the discriminant can also be expressed as

$$\Delta = (c_4^3 - c_6^2)/1728$$

(notice that $1728 = 2^6 3^3$). A curve is then non-singular if and only if $\Delta \neq 0$.

When $\Delta \neq 0$, the *j-invariant* of the curve is defined as

$$j(E) = c_4^3/\Delta. \tag{III.5}$$

The j-invariant is closely related to the notion of *elliptic curve isomorphism.* Two elliptic curves defined by Weierstrass equations E (with variables X, Y) and E' (with variables X', Y') are isomorphic over K if and only if there exist constants $r, s, t \in K$ and $u \in K^*$, such that the change of variables

$$X = u^2 X' + r\,, \quad Y = u^3 Y' + s u^2 X' + t \tag{III.6}$$

transforms E into E'. The transformation in Equations (III.6) is referred to as an *admissible change of variables.* Clearly, this transformation is reversible, and its inverse also defines an admissible change of variables that transforms E' into E. Such an isomorphism defines a bijection between the set of rational points in E and the set of rational points in E'. Notice that isomorphism is defined relative to the field K. Curves that are not isomorphic over K can become so over an extension $\hat{K} \supseteq K$.

Curve isomorphism is an equivalence relation. The following lemma establishes the fact that, *over the algebraic closure* $\overline{K}$, the j-invariant characterizes the equivalence classes in this relation. Proofs for all characteristics can be found in [**147**].

LEMMA III.1. *Two elliptic curves that are isomorphic over K have the same j-invariant. Conversely, two curves with the same j-invariant are isomorphic over $\overline{K}$.*

III.2. The Group Law

Assume, for the moment, that $\mathrm{char}(K) \neq 2, 3$, and consider the admissible change of variables given by

$$X = X' - \frac{b_2}{12}, \quad Y = Y' - \frac{a_1}{2}(X' - \frac{b_2}{12}) - \frac{a_3}{2},$$

with b_2 defined as in Equations (III.4). This change of variables transforms the long Weierstrass form in Equation (III.3) to the equation of an isomorphic curve given in the *short Weierstrass form,*

$$E : Y^2 = X^3 + aX + b, \tag{III.7}$$

for some $a, b \in K$.

Let P and Q be two distinct rational points on E. The straight line joining P and Q must intersect the curve at one further point, say R, since we are intersecting a line with a cubic curve. The point R will also be rational since the line, the curve and the points P and Q are themselves all defined over K.

FIGURE III.1. Adding two points on an elliptic curve

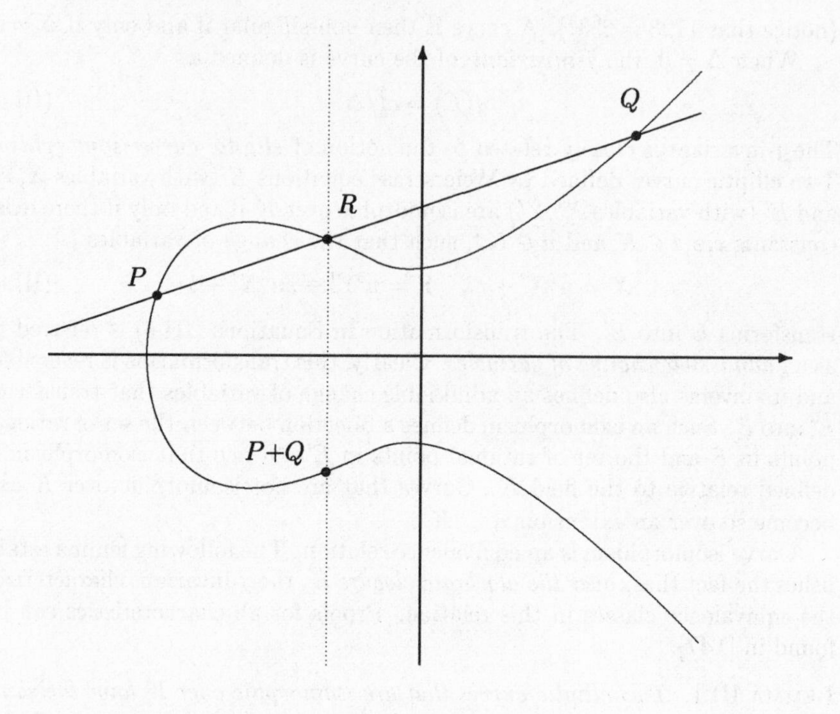

If we then reflect R in the x-axis we obtain another rational point which we shall call $P+Q$ (see Figure III.1 for a visualization over the reals).

To add P to itself, or to *double* P in the jargon, we take the tangent to the curve at P. Such a line must intersect $E(K)$ in exactly one other point, say R, as E is defined by a cubic equation. Again we reflect R in the x-axis to obtain a point which we call $[2]P = P + P$ (see Figure III.2). If the tangent to the point is vertical, it 'intersects' the curve at the point at infinity and $P + P = \mathcal{O}$, i.e., P is a point of order 2.

The above process of determining $P+Q$ given P and Q is often called the *chord–tangent process.* The operation on points which we have just explained can be shown to define an additive abelian group law on $E(\hat{K})$, for any field $K \subseteq \hat{K} \subseteq \overline{K}$, with the point at infinity, $\mathcal{O}$, as the zero. The whole law can be summarized in the statement that three points on the curve will sum to zero if and only if they lie on a straight line.

FIGURE III.2. Doubling a point on an elliptic curve

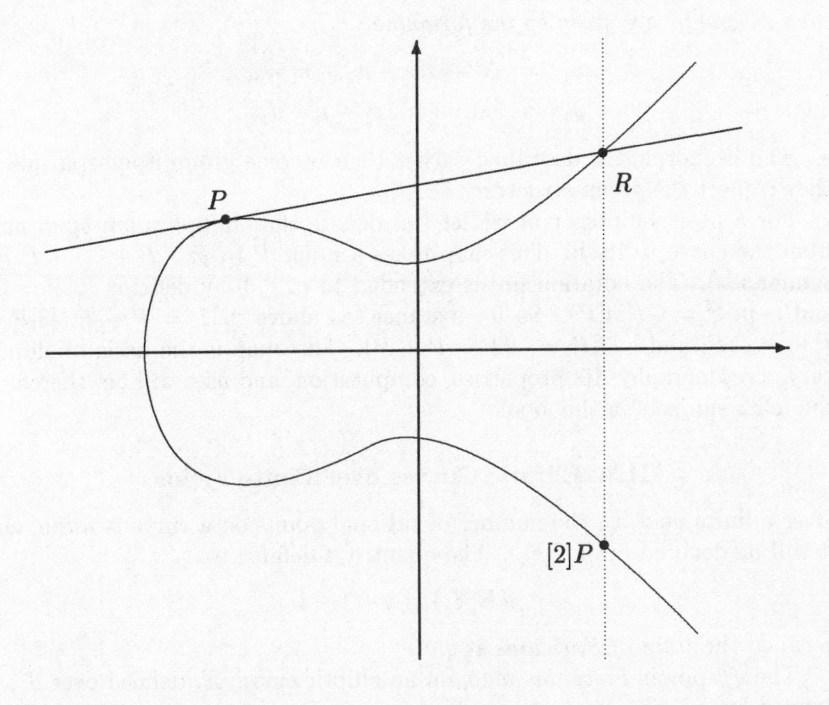

Using this geometric definition, which is readily extended to $\mathrm{char}(K) = 2$ or 3, we can determine explicit algebraic formulae for the above group law. These formulae are valid in any characteristic.

LEMMA III.2. *Let E denote an elliptic curve given by*

$$E : Y^2 + a_1XY + a_3Y = X^3 + a_2X^2 + a_4X + a_6$$

and let $P_1 = (x_1, y_1)$ and $P_2 = (x_2, y_2)$ denote points on the curve. Then

$$-P_1 = (x_1, -y_1 - a_1x_1 - a_3).$$

Set

$$\lambda = \frac{y_2 - y_1}{x_2 - x_1}\ ,\ \mu = \frac{y_1x_2 - y_2x_1}{x_2 - x_1}$$

when $x_1 \neq x_2$, and set

$$\lambda = \frac{3x_1^2 + 2a_2x_1 + a_4 - a_1y_1}{2y_1 + a_1x_1 + a_3}\ ,\ \mu = \frac{-x_1^3 + a_4x_1 + 2a_6 - a_3y_1}{2y_1 + a_1x_1 + a_3}$$

when $x_1 = x_2$ *and* $P_2 \neq -P_1$. *If*

$$P_3 = (x_3, y_3) = P_1 + P_2 \neq \mathcal{O}$$

then x_3 *and* y_3 *are given by the formulae*

$$\begin{aligned} x_3 &= \lambda^2 + a_1\lambda - a_2 - x_1 - x_2, \\ y_3 &= -(\lambda + a_1)x_3 - \mu - a_3. \end{aligned}$$

The isomorphisms described earlier then become group isomorphisms as they respect the group structure.

For a positive integer m we let $[m]$ denote the *multiplication-by-m* map from the curve to itself. This map takes a point P to $P + P + \cdots + P$ (m summands). The notation $[m]$ is extended to $m \leq 0$ by defining $[0]P = \mathcal{O}$, and $[-m]P = -([m]P)$. So for instance, as above, $[2]P = P + P$, $[3]P = P + P + P$, and $[-3]P = -(P + P + P)$. This map is the basis of elliptic curve cryptography. Its properties, computation, and uses will be, therefore, the main subjects in this book.

III.3. Elliptic Curves over Finite Fields

Over a finite field $\mathbb{F}_q$, the number of rational points on a curve is finite, and it will be denoted by $\#E(\mathbb{F}_q)$. The quantity t defined by

$$\#E(\mathbb{F}_q) = q + 1 - t$$

is called the *trace of Frobenius* at q.

The qth-power *Frobenius map*, on an elliptic curve, E, defined over $\mathbb{F}_q$, is defined by

$$\varphi : \begin{cases} E(\overline{\mathbb{F}}_q) & \longrightarrow & E(\overline{\mathbb{F}}_q) \\ (x, y) & \longmapsto & (x^q, y^q), \\ \mathcal{O} & \longmapsto & \mathcal{O}. \end{cases}$$

It is readily verified that φ maps points on E to points on E, and that it respects the group law. In other words the map φ is a group endomorphism of E over $\mathbb{F}_q$, referred to as the *Frobenius endomorphism*.

The trace of Frobenius t and the Frobenius endomorphism φ will play a fundamental role in our study of elliptic curves. They are linked by the equation

$$\varphi^2 - [t]\varphi + [q] = [0],$$

that is, for any point $P = (x, y)$ on the curve, we have

$$(x^{q^2}, y^{q^2}) - [t](x^q, y^q) + [q](x, y) = \mathcal{O},$$

where addition and subtraction denote curve operations.

A first approximation to the order of $E(\mathbb{F}_q)$ is given by the following well known theorem of Hasse, a proof of which can be found in [**147**, Theorem V.1.1].

THEOREM III.3 (H. Hasse, 1933). *The trace of Frobenius satisfies*

$$|t| \leq 2\sqrt{q}.$$

By Hasse's Theorem, the number of points on the curve, for large values of q, is in a narrow range of width $4\sqrt{q}$ about the value $q+1$. To intuitively see why this should be so, notice that about half of all the q possible x-coordinates in $\mathbb{F}_q$ will give rise to a solution y. All but at most three of these will have two corresponding y-coordinates, the exceptions being the points of order two (i.e. those points with y-coordinate equal to zero in the short Weierstrass form of the curve). To this expected number q of rational points we add the point at infinity making a total of $q+1$ expected rational points on a curve over $\mathbb{F}_q$.

This observation tells us how to choose elements of $E(\mathbb{F}_q)$ with an (almost) uniform distribution.

ALGORITHM III.1: **Determine a Random Point in $E(\mathbb{F}_q)$.**

```
INPUT:  An elliptic curve E(F_q).
OUTPUT: A 'random' point P ∈ E(F_q).
1.  Do
2.        Pick a random x ∈ F_q.
3.        Substitute x for X in Equation (III.3).
4.        Attempt to solve the resulting quadratic equation in Y,
          using the techniques in Sections II.1.5 and II.2.4.
5.        If solutions y are found, flip a coin to decide
          which y to choose and set P = (x,y).
6.  Until a point P is found.
7.  Return P.
```

For curves over $\mathbb{F}_p$, where p is a prime, there is an elliptic curve with group of rational points of any given order in the interval $(p+1-2\sqrt{p}, p+1+2\sqrt{p})$. In the sub-interval $(p+1-\sqrt{p}, p+1+\sqrt{p})$ each order occurs with an almost uniform distribution. This fact is the basis behind the ECM factoring algorithm of Lenstra (see [**78**] and Section IX.1). However, this distribution has some very subtle properties; see [**89**] for details. Over fields of characteristic two the statement is not true.

There are two particular classes of curves which, under certain conditions, will prove to be cryptographically weak: anomalous curves and supersingular curves. The curve $E(\mathbb{F}_q)$ is said to be *anomalous* if its trace of Frobenius is 1, giving $\#E(\mathbb{F}_q) = q$. These curves are weak when $q = p$, the field characteristic. The attack against such curves is discussed in Chapter V. The curve $E(\mathbb{F}_q)$ is said to be *supersingular* if the characteristic p divides the trace of Frobenius, t. Equivalently, it can be shown that a curve over $\mathbb{F}_q$ with characteristic p is supersingular if and only if (i) $p = 2$ or 3 and $j(E) = 0$ or

(ii) $p \geq 5$ and $t = 0$. The MOV attack on general elliptic curves, described in Chapter V, is particularly effective for supersingular curves, rendering them unsuitable for cryptographic purposes.

Contrary to the case of elliptic curves over $\mathbb{Q}$, where the characterization of possible ranks of groups $E(\mathbb{Q})$ is an open problem, this rank is well characterized over finite fields, where we have

$$E(\mathbb{F}_q) \cong (\mathbb{Z}/d_1\mathbb{Z}) \times (\mathbb{Z}/d_2\mathbb{Z}).$$

Here, by the structure theorem for finite abelian groups, d_1 divides both d_2 and $q-1$, and we include the case $d_1 = 1$.

As was apparent from the earlier discussion, the cases $\mathrm{char}(K) = 2, 3$ often require separate treatment. Practical implementations of elliptic curve cryptosystems are usually based on either $\mathbb{F}_{2^n}$, i.e., characteristic two, or $\mathbb{F}_p$ for large primes p. Therefore, the remainder of this book will focus on fields of characteristics two and $p > 3$, and will omit the separate treatment of the case $\mathrm{char}(K) = 3$. Most arguments, though, carry easily to characteristic three, with modifications that are well documented in the literature.

III.3.1. Curves in fields of characteristic $p > 3$. Assume $K = \mathbb{F}_q$, where $q = p^n$ for a prime $p > 3$ and an integer $n \geq 1$. As mentioned, the curve equation in this case can be simplified to the short Weierstrass form

$$E_{a,b} : Y^2 = X^3 + aX + b.$$

The discriminant of the curve then reduces to $\Delta = -16(4a^3 + 27b^2)$, and its j-invariant to $j(E) = -1728(4a)^3/\Delta$. The isomorphism classes of curves over K in this case are characterized by the relation

$$E_{a,b} \cong E_{a',b'} \text{ if and only if } a' = u^4 a,\ b' = u^6 b,$$

for some $u \in K^*$.

The formulae for the group law in Lemma III.2 simplify to

$$-P_1 = (x_1, -y_1).$$

When $x_1 \neq x_2$ we set

$$\lambda = \frac{y_2 - y_1}{x_2 - x_1},$$

and when $x_1 = x_2$, $y_1 \neq 0$ we set

$$\lambda = \frac{3x_1^2 + a}{2y_1}.$$

If

$$P_3 = (x_3, y_3) = P_1 + P_2 \neq \mathcal{O},$$

then x_3 and y_3 are given by the formulae

$$\begin{aligned} x_3 &= \lambda^2 - x_1 - x_2, \\ y_3 &= (x_1 - x_3)\lambda - y_1. \end{aligned}$$

Write $g(X) = X^3 + aX + b$, so that the curve equation is $Y^2 = g(X)$. The rational points of order two on the curve are of the form $(\xi, 0)$, where ξ is a zero of $g(X)$ in K. The polynomial $g(X)$ can have zero, one, or three such zeros. All other values of X for which $g(X)$ is a quadratic residue in K yield two distinct points on the curve. Therefore, counting also the point $\mathcal{O}$, we have $\#E(K) \equiv s \pmod 2$, where $s = 1$ if g is irreducible over K, 0 otherwise.

A *twist* of a curve given in short Weierstrass form $E_{a,b}$ is given by $E_{a',b'}$ where $a' = v^2 a$, $b' = v^3 b$ for some quadratic non-residue $v \in K$. By the characterization of isomorphism classes above, the twist is unique up to isomorphisms over K, and it is itself isomorphic to the original curve, over $\overline{K}$ (in fact, it is so over $\mathbb{F}_{q^2}$, where v becomes a quadratic residue). The orders of the groups of rational points of the two curves satisfy the relation

$$\#E_{a,b}(K) + \#E_{a',b'}(K) = 2q + 2.$$

To verify this, write $g_v(X) = v^3 g(X/v)$, so that we have $E_{a',b'} : Y^2 = g_v(X)$. For $x \in K$, if $g_v(x) = 0$ then $g(x/v) = 0$, contributing a single point to each of the curves. If $g_v(x)$ is a non-zero quadratic residue, then $g(x/v) = g_v(x)/v^3$ is a non-residue; $E_{a',b'}$ gets two points, $E_{a,b}$ gets none. Similarly, if $g_v(x)$ is a non-residue, then $E_{a,b}$ gets two points, $E_{a',b'}$ gets none. Hence, each element of K contributes two counts to the sum $\#E_{a,b}(K) + \#E_{a',b'}(K)$, giving, together with the point at infinity counted twice, a total of $2q + 2$ points.

This property of the twist is useful when searching for 'good' curves in cryptography, where it is required to determine the order of the group of rational points. This is a computationally intensive problem, which we deal with extensively in Chapters VI, VII and VIII. Once the group order has been determined for a curve, its determination for the twist is straightforward. Thus, we get the orders of two groups 'for the price of one'.

III.3.2. Curves in fields of characteristic two. We now specialize to the case where $q = 2^n$, $n \geq 1$. In this case, the expression for the j-invariant reduces to $j(E) = a_1^{12}/\Delta$. In characteristic two, the condition $j(E) = 0$, i.e. $a_1 = 0$, is equivalent to the curve being supersingular. As mentioned, this very special type of curve is avoided in cryptography (see details on the MOV attack in Chapter V). We assume, therefore, that $j(E) \neq 0$.

Under these assumptions, a representative for each isomorphism class of elliptic curves over $\mathbb{F}_q$ is given by [**147**]:

$$E_{a_2,a_6} : Y^2 + XY = X^3 + a_2 X^2 + a_6, \tag{III.8}$$

where $a_6 \in \mathbb{F}_q^*$ and $a_2 \in \{0, \gamma\}$ with γ a fixed element in $\mathbb{F}_q$ of trace $\mathrm{Tr}_{q|2}(\gamma) = 1$. We recall from Chapter II that $\mathrm{Tr}_{q|2}$ is the linear trace from $\mathbb{F}_q$ to $\mathbb{F}_2$. This function is not directly related to the trace of Frobenius, and no confusion should arise since they are used in quite different contexts.

The formulae for the group law in Lemma III.2 then simplify to

$$-P_1 = (x_1, y_1 + x_1).$$

When $x_1 \neq x_2$ we set

$$\lambda = \frac{y_2 + y_1}{x_2 + x_1}, \ \mu = \frac{y_1 x_2 + y_2 x_1}{x_2 + x_1}$$

and when $x_1 = x_2 \neq 0$ we set

$$\lambda = \frac{x_1^2 + y_1}{x_1}, \ \mu = x_1^2.$$

If

$$P_3 = (x_3, y_3) = P_1 + P_2 \neq \mathcal{O},$$

then x_3 and y_3 are given by the formulae

$$\begin{aligned} x_3 &= \lambda^2 + \lambda + a_2 + x_1 + x_2, \\ y_3 &= (\lambda + 1)x_3 + \mu \\ &= (x_1 + x_3)\lambda + x_3 + y_1. \end{aligned}$$

The following lemma restricts the possible values of $\#E_{a_2,a_6}(\mathbb{F}_q)$ as a function of the isomorphism class. Recall that each element $a \in \mathbb{F}_q$ has a unique square root, $\sqrt{a} = a^{q/2}$, in the field.

LEMMA III.4. *Consider an elliptic curve defined by Equation (III.8) over $\mathbb{F}_q$, $q = 2^n$. Then,*

$$\#E_{a_2,a_6}(\mathbb{F}_q) \equiv \begin{cases} 0 \pmod 4 & \text{if } \mathrm{Tr}_{q|2}(a_2) = 0, \\ 2 \pmod 4 & \text{if } \mathrm{Tr}_{q|2}(a_2) = 1. \end{cases}$$

PROOF. Setting $X = 0$ yields $(0, \sqrt{a_6})$, the unique point of order two on the curve. To count points with $X \neq 0$, we divide Equation (III.8) by X^2, and write $U = Y/X$, obtaining the equivalent equation

$$U^2 + U = X + a_2 + \frac{a_6}{X^2}.$$

It is well known (see, e.g., [86]) that, for a given $X \in \mathbb{F}_q^*$, this quadratic equation has two distinct solutions U and $U+1$ in $\mathbb{F}_q$ if and only if $\mathrm{Tr}_{q|2}(X + a_2 + a_6/X^2) = 0$ or, equivalently, $\mathrm{Tr}_{q|2}(a_2) = \mathrm{Tr}_{q|2}(X^2 + a_6/X^2)$. If X satisfies this equality, so does $\sqrt{a_6}/X$. These two values are different whenever $X \neq \sqrt[4]{a_6}$. Hence, the values of X in $\mathbb{F}_q^* - \{\sqrt[4]{a_6}\}$ contribute a number of points divisible by four to $\#E_{a_2,a_6}(\mathbb{F}_q)$. When $\mathrm{Tr}_{q|2}(a_2) = 0$, $X = \sqrt[4]{a_6}$ contributes two more points. Counting also the points $(0, \sqrt{a_6})$ and $\mathcal{O}$ yields the result of the lemma. □

For a given value of a_6, the two curves E_{0,a_6} and E_{γ,a_6} are twists of each other and their orders satisfy the relation

$$\#E_{0,a_6}(\mathbb{F}_q) + \#E_{\gamma,a_6}(\mathbb{F}_q) = 2q + 2.$$

This is verified by inspecting the proof of Lemma III.4: each value of $X \in \mathbb{F}_q^*$ contributes two points to exactly one of the curves, for a total of $2q - 2$ points.

In addition, the points $(0, \sqrt{a_6})$ and $\mathcal{O}$ are common to both curves and are counted twice in the sum, bringing the total up to $2q+2$.

Similarly to the case of odd characteristic, the curves E_{0,a_6} and E_{γ,a_6} are non-isomorphic over $\mathbb{F}_q$, but are isomorphic over $\mathbb{F}_{q^2}$, as $\mathrm{Tr}_{q^2|2}(\gamma) = 0$ for all $\gamma \in \mathbb{F}_q$.

III.4. The Division Polynomials

The division polynomials are of fundamental importance in Schoof's algorithm for computing the number of points on an elliptic curve over a finite field, the subject of Chapter VII. In this section we define these polynomials and discuss some of their basic properties. References for much of the following are [**147**] and [**72**]. The specific formulae for the division polynomials in the general case follow [**81**] and [**85**].

From inspection of the algebraic expressions for the group law given in Section III.2, it is clear that the coordinates of the sum P_1+P_2 of two points on the curve are rational functions of the coordinates of P_1 and P_2. By repeated application of the formulae, it follows that the multiplication-by-m map

$$(x,y) \mapsto [m](x,y)$$

can be expressed in terms of rational functions in x and y. More specifically, we have the following result.

LEMMA III.5. *Let E be an elliptic curve defined over a field K, and let m be a positive integer. There exist polynomials $\psi_m, \theta_m, \omega_m \in K[x,y]$ such that, for $P=(x,y) \in E(\overline{K})$ such that $[m]P \neq \mathcal{O}$, we have*

$$[m]P = \left(\frac{\theta_m(x,y)}{\psi_m(x,y)^2}, \frac{\omega_m(x,y)}{\psi_m(x,y)^3} \right). \tag{III.9}$$

The polynomial $\psi_m(x,y)$ is called the *mth division polynomial* of the curve E. As will be shown below, the sequences θ_m and ω_m can be expressed in terms of the sequence ψ_m.

We now present explicit (recursive) formulae for the polynomials ψ_m, θ_m and ω_m. Consider the general Weierstrass equation E of the elliptic curve over K given in Equation (III.3), and the constants derived from the curve parameters given in Equations (III.4). The mth division polynomial $\psi_m(x,y)$, $m \geq 0$, is defined by the following recursion, in which we suppress the variables:

$$\psi_0 = 0, \ \psi_1 = 1,$$
$$\psi_2 = 2y + a_1 x + a_3,$$
$$\psi_3 = 3x^4 + b_2 x^3 + 3b_4 x^2 + 3b_6 x + b_8,$$
$$\psi_4 = \left(2x^6 + b_2 x^5 + 5b_4 x^4 + 10 b_6 x^3 + 10 b_8 x^2 + (b_2 b_8 - b_4 b_6)x + b_4 b_8 - b_6^2\right)\psi_2,$$
$$\psi_{2m+1} = \psi_{m+2}\psi_m^3 - \psi_{m-1}\psi_{m+1}^3, \ m \geq 2,$$

and

$$\psi_{2m} = \frac{\left(\psi_{m+2}\psi_{m-1}^2 - \psi_{m-2}\psi_{m+1}^2\right)\psi_m}{\psi_2}, \quad m > 2.$$

It can be shown, by induction, that the numerator in the expression for ψ_{2m} is divisible by ψ_2^2. Therefore, ψ_{2m}, $m \geq 1$, is a polynomial divisible by ψ_2. Since the division polynomials will always be evaluated at points on the curve, the computation of ψ_m can be carried out modulo the equation of the curve. In particular, we can assume that the degree of ψ_m in y never exceeds one. This reduction will be implicitly assumed in the sequel when dealing with the polynomials ψ_m. With the ψ_m computed according to the above recursion, the polynomials θ_m are given by

$$\theta_m = x\psi_m^2 - \psi_{m-1}\psi_{m+1}, \quad m \geq 1,$$

and, when $\mathrm{char}(K) \neq 2$, the polynomials ω_m are defined by

$$2\psi_m\omega_m = \psi_{2m} - (a_1\theta_m + a_3\psi_m^2)\psi_m^2, \quad m \geq 1.$$

With the given recursion for the polynomials ψ_m, and the formulae for θ_m and ω_m, Lemma III.5 follows directly from the formulae for the group law, and some symbolic manipulation dexterity. In the case of characteristic two, the ω_m require a slightly different treatment. This case will be addressed, for non-supersingular curves, in Section III.4.2. Expressions for the supersingular case can be found in [**64**].

When K is the finite field $\mathbb{F}_q$, $E(\overline{K})$ is a *torsion group*, that is, every point P on the curve E has finite order. For a non-negative integer m, the set of *m-torsion points* of E, denoted by $E[m]$, is defined by

$$E[m] = \{\, P \in E(\overline{K}) \mid [m]P = \mathcal{O} \,\}.$$

It is readily verified that $E[m]$ is a subgroup of $E(\overline{K})$. When we are interested in K-rational points in $E[m]$, we will use the notation $E(K)[m] = E(K) \cap E[m]$. Thus, $E(\overline{K})[m] = E[m]$. Clearly, $E(K)[m] \subseteq E[m] \subseteq E(\overline{K})$, where inclusion is interpreted as the subgroup relation.

By definition, $\mathcal{O} \in E[m]$ for all m. The mth division polynomial ψ_m characterizes the other m-torsion points on E, as stated in the following theorem.

THEOREM III.6. *Let P be a point in $E(\overline{K}) \setminus \{\mathcal{O}\}$, and let $m \geq 1$. Then, $P \in E[m]$ if and only if $\psi_m(P) = 0$.*

It turns out that the characterization of m-torsion points can be achieved with univariate polynomials derived from the bivariate ψ_m. Define

$$\overline{f}_m = \begin{cases} \psi_m, & m \text{ odd}, \\ \psi_m/\psi_2, & m \text{ even}. \end{cases}$$

By observing that y enters into the recursion for the ψ_m only through the polynomial ψ_2, and that $\psi_2^2 \mod E$ does not depend on y, it is readily verified that $\overline{f}_m$ is a polynomial that depends only on x. The degree of $\overline{f}_m$ is at most $(m^2-1)/2$ if m is odd, and at most $(m^2-4)/2$ if m is even (the degrees are exact if $\mathrm{char}(K)$ does not divide m for m odd, or $m/2$ for m even). Theorem III.6 can now be recast in terms of the polynomials $\overline{f}_m$.

COROLLARY III.7. *Let $P = (x, y)$ be a point in $E(\overline{K}) - \{\mathcal{O}\}$, such that $[2]P \neq \mathcal{O}$, and let $m \geq 2$. Then, $P \in E[m]$ if and only if $\overline{f}_m(x) = 0$.*

Corollary III.7 excludes 2-torsion points. These points satisfy $\psi_2(P) = 0$, the part that was divided out of ψ_m to obtain $\overline{f}_m$ when m is even.

Let $F(x) = 4x^3 + b_2x^2 + 2b_4x + b_6$. The polynomials $\overline{f}_m$ satisfy the following recursion, where variables are again omitted, and ψ_2, ψ_3 and ψ_4 are as defined before:

$$\overline{f}_0 = 0,\ \ \overline{f}_1 = 1,\ \ \overline{f}_2 = 1,\ \ \overline{f}_3 = \psi_3,\ \ \overline{f}_4 = \psi_4/\psi_2,$$

$$\begin{aligned}
\overline{f}_{2m+1} &= \begin{cases} \overline{f}_{m+2}\overline{f}_m^3 - F^2\overline{f}_{m-1}\overline{f}_{m+1}^3, & m \text{ odd}, m \geq 3, \\ F^2\overline{f}_{m+2}\overline{f}_m^3 - \overline{f}_{m-1}\overline{f}_{m+1}^3, & m \text{ even}, m \geq 2, \end{cases} \\
\overline{f}_{2m} &= (\overline{f}_{m+2}\overline{f}_{m-1}^2 - \overline{f}_{m-2}\overline{f}_{m+1}^2)\overline{f}_m, \ m > 2.
\end{aligned}$$

Our interest in this book will involve the two cases $\mathrm{char}(K) > 3$ and $\mathrm{char}(K) = 2$. The above discussion is specialized to these two cases, in turn.

III.4.1. Characteristic $p > 3$.

For this case the curve equation can be assumed in the form

$$Y^2 = X^3 + aX + b, \quad a, b \in \mathbb{F}_p,$$

and so, in the above formulae for the polynomials ψ_m and $\overline{f}_m$, we have $a_1 = a_2 = a_3 = 0$, $a_4 = a$, $a_6 = b$, $b_2 = 0$, $b_4 = 2a$, $b_6 = 4b$, and $b_8 = -a^2$. The recursion for ψ_m then simplifies to

$$\begin{aligned}
\psi_0 &= 0, \\
\psi_1 &= 1, \\
\psi_2 &= 2y, \\
\psi_3 &= 3x^4 + 6ax^2 + 12bx - a^2, \\
\psi_4 &= 4y(x^6 + 5ax^4 + 20bx^3 - 5a^2x^2 - 4abx - 8b^2 - a^3), \\
\psi_{2m+1} &= \psi_{m+2}\psi_m^3 - \psi_{m-1}\psi_{m+1}^3, \quad m \geq 2, \\
\psi_{2m} &= (\psi_{m+2}\psi_{m-1}^2 - \psi_{m-2}\psi_{m+1}^2)\psi_m/2y, \quad m > 2.
\end{aligned}$$

For an integer $m \geq 2$, and a point $P = (x, y) \in E(\overline{K}) \setminus E[m]$, Lemma III.5 takes the form

$$[m]P = \left(x - \frac{\psi_{m-1}\psi_{m+1}}{\psi_m^2}, \ \frac{\psi_{m+2}\psi_{m-1}^2 - \psi_{m-2}\psi_{m+1}^2}{4y\psi_m^3} \right),$$

where $\psi_m = \psi_m(x, y)$. This formula is easily cast in terms of the univariate polynomials $\overline{f}_m$, by noting that for the particular form of the curve equation under consideration, we have $\psi_m = 2y\overline{f}_m$ when m is even, $\psi_m = \overline{f}_m$ when m is odd. The recursions for the $\overline{f}_m$ are as in the general case, with $F(x) = 4(x^3 + ax + b)$ (which is equal to $4y^2$ modulo the curve equation).

III.4.2. Characteristic two. We consider only non-supersingular curves, defined by equations of the form

$$Y^2 + XY = X^3 + a_2X^2 + a_6.$$

Thus, we have $a_1 = 1$, $a_3 = a_4 = 0$, and consequently $b_2 = 1$, $b_4 = b_6 = 0$, $b_8 = a_6$. The recursion for the polynomials ψ_m simplifies to

$$\begin{aligned}
\psi_0 &= 0,\\
\psi_1 &= 1,\\
\psi_2 &= x,\\
\psi_3 &= x^4 + x^3 + a_6,\\
\psi_4 &= x^6 + a_6x^2,\\
\psi_{2m+1} &= \psi_{m+2}\psi_m^3 + \psi_{m-1}\psi_{m+1}^3,\ m \geq 2,\\
\psi_{2m} &= (\psi_{m+2}\psi_{m-1}^2 + \psi_{m-2}\psi_{m+1}^2)\psi_m/x,\ m > 3.
\end{aligned}$$

We observe that, with this recursion, all the ψ_m are polynomials in x only. We shall emphasize this fact by defining, in this case, $f_m(x) = \psi_m(x, y)$. The formulae for the mapping $[m]$ then take the form

$$[m]P = \left(x + \frac{f_{m-1}f_{m+1}}{f_m^2}, x + y + \frac{(x^2 + x + y)f_{m-1}f_mf_{m+1} + f_{m-2}f_{m+1}^2}{xf_m^3} \right),$$

for $m \geq 2$ and points $P = (x, y) \in E(\overline{K}) \setminus E[m]$. The polynomials $\overline{f}_m$ defined in the general case satisfy, in this case, $x\overline{f}_m = f_m$ when m is even, $\overline{f}_m = f_m$ otherwise. In fact, in our description of point counting algorithms in Chapter VII, we shall use mostly the polynomials f_m, a notation which is extended by defining $f_m = \overline{f}_m$ for all m in the odd characteristic case.

Formally, the polynomials ψ_m are called the division polynomials. However, in the cases of interest here, the similar role of the univariate polynomials f_m will justify our referring to these also as division polynomials.

III.5. The Weil Pairing

Let E denote an elliptic curve over a field K, with $E[m]$ its group of m-torsion points. It can be shown that there are m^2 such points in the case $\gcd(m, p) = 1$, where p is the characteristic of the field. The structure of the m-torsion group of an elliptic curve is determined by the following result:

LEMMA III.8. *Let E be an elliptic curve over K and let char$(K) = p$ and $m \in \mathbb{Z}_{>0}$.*

- *If $p = 0$ or p does not divide m then*

$$E[m] \cong (\mathbb{Z}/m\mathbb{Z}) \times (\mathbb{Z}/m\mathbb{Z}).$$

- *If* $p > 0$ *then*
$$E[p^r] \cong (\mathbb{Z}/p^r\mathbb{Z}) \text{ or } \{0\}.$$

Another important fact about the m-torsion structure of an elliptic curve over a finite field, which will be required in a later chapter, is given by

LEMMA III.9 ([**8**]). *Let E denote an elliptic curve over $\mathbb{F}_q$, and suppose that m is a prime which divides $\#E(\mathbb{F}_q)$ but which does not divide $q-1$ and is not equal to the characteristic of $\mathbb{F}_q$. Then $E(\mathbb{F}_{q^k})$ contains the m^2 points of order m if and only if m divides $q^k - 1$.*

We now let $m \in \mathbb{Z}_{\geq 2}$ denote an integer, coprime to the characteristic of K if $\mathrm{char}(K) > 0$. The *Weil pairing* [**147**] is a function

$$E[m] \times E[m] \longrightarrow \mu_m,$$

where μ_m is the group of mth roots of unity in $\overline{K}$, which occurs throughout the theory of elliptic curves. We can define the Weil pairing as follows. Let $S, T \in E[m]$ and choose a function g on E whose divisor satisfies

$$\mathrm{div}(g) = \sum_{R \in E[m]} (T' + R) - (R),$$

with $T' \in E(\overline{K})$ such that $[m]T' = T$. Then

$$e_m : \begin{cases} E[m] \times E[m] & \longrightarrow & \mu_m \\ (S, T) & \longmapsto & \dfrac{g(X+S)}{g(X)} \end{cases}$$

for any point $X \in E(K)$ for which g is both defined and non-zero at X and $X + S$. It can then be shown that the following holds.

LEMMA III.10. *The Weil pairing is a bilinear, alternating, non-degenerate pairing which is Galois equivariant. In other words,*

$$\begin{aligned} e_m(S_1 + S_2, T) &= e_m(S_1, T) e_m(S_2, T), \\ e_m(S, T_1 + T_2) &= e_m(S, T_1) e_m(S, T_2), \\ e_m(S, T) &= e_m(T, S)^{-1}, \\ e_m(S, T) &= 1 \text{ for all } S \text{ if and only if } T = \mathcal{O}, \\ e_m(S^\sigma, T^\sigma) &= e_m(S, T)^\sigma \text{ for all } \sigma \in \mathrm{Gal}(\overline{K}/K). \end{aligned}$$

There is another definition of the Weil pairing which makes it easier to compute. We let P and Q denote two elements of $E[m]$ and let A, B denote divisors of degree zero such that A and B have disjoint support and

$$A \sim (P) - (\mathcal{O}), \; B \sim (Q) - (\mathcal{O}).$$

In practice we choose points $T, U \in E$ such that $P + T \neq U$, $P + T \neq Q + U$, $T \neq U$ and $T \neq Q + U$. We then see that $A = (P + T) - (T)$ and $B = (Q + U) - (U)$ satisfy our requirements.

We then let f_A and f_B denote two functions whose divisors are mA and mB respectively. The Weil pairing can then be defined by

$$e_m(P,Q) = f_A(B)/f_B(A),$$

which, owing to our choice of A and B, becomes

$$e_m(P,Q) = \frac{f_A(Q+U)f_B(T)}{f_A(U)f_B(P+T)}.$$

So all that remains is to compute f_A and f_B. This can be done by a method of Miller which is explained in [**97**] and [**98**]. One has to be careful that the functions one produces are defined and are non-zero at the relevant points, but by careful choice of T and U this can be accomplished with no problem.

III.6. Isogenies, Endomorphisms and Torsion

Let E_1 and E_2 be elliptic curves defined over a field K, with respective function fields $\overline{K}(E_1)$ and $\overline{K}(E_2)$. A *morphism* from E_1 to E_2 is a rational map which is regular at every point of E_1. A non-constant morphism, ϕ, which maps the identity element on E_1 to the identity element on E_2 is called an *isogeny*,

$$\phi : E_1 \longrightarrow E_2.$$

The map which sends every point on E_1 to $\mathcal{O}$ on E_2 is also called an isogeny. It is the zero isogeny, and is the only constant isogeny. Isogenies play a crucial role in the theory of elliptic curves. In this section we summarize the main results that will be required later.

Suppose that the isogeny ϕ is non-constant, i.e. $\phi(E_1) \neq \{\mathcal{O}\}$. Then, ϕ induces an injection of function fields which fixes $\overline{K}$,

$$\phi^* : \begin{cases} \overline{K}(E_2) & \longrightarrow & \overline{K}(E_1) \\ f & \longmapsto & f \circ \phi. \end{cases}$$

We say that the isogeny is separable, inseparable or purely inseparable if the corresponding extension of function fields, $\overline{K}(E_1)/\phi^*\overline{K}(E_2)$ is separable, inseparable or purely inseparable. If ϕ is constant we define its degree to be zero, otherwise we define its degree by

$$\deg \phi = [\overline{K}(E_1) : \phi^*\overline{K}(E_2)].$$

Every non-constant isogeny ϕ is surjective over $\overline{K}$, that is $\phi(E_1) = E_2$. An isogeny is always a group homomorphism, and the kernel of a non-constant isogeny ϕ is always a finite subgroup of $E_1(\overline{K})$, usually denoted by $E[\phi]$. The degree n of a separable isogeny ϕ is equal to its degree as a finite map of curves and is hence equal to the size of $E[\phi]$.

The simplest example of a separable isogeny is the multiplication-by-m map, $[m]$, from a curve to itself. If K is a finite field $\mathbb{F}_q$ and E is an elliptic curve defined over K, then the simplest example of a purely inseparable isogeny is the Frobenius endomorphism φ. If E is an elliptic curve over $\mathbb{F}_q$

with N points then the isogenies $[1]$, $[N+1]$ and φ are identical as maps on $E(\mathbb{F}_q)$. However, they are all different when considered over the algebraic closure of $\mathbb{F}_q$.

Some basic facts about isogenies are

THEOREM III.11 (Theorem 11.66 of [**60**]). *Let E denote an elliptic curve defined over a field K and let S denote a finite subgroup of E which is Galois stable over K. Then there exist an elliptic curve E', also defined over K, and a unique separable isogeny $\phi : E \longrightarrow E'$ with kernel equal to S.*

When $K = \mathbb{F}_q$, the subgroup S in Theorem III.11 is Galois stable if and only if it is closed under the operation of the Frobenius map. Also, the notation E/S is often used for the curve E' described in the theorem, that is,

$$\phi : E \longrightarrow E/S.$$

This notation is obvious from a group-theoretic point of view, but it also conveys the less obvious fact that the quotient group E/S corresponds to the group of points of an elliptic curve.

To every non-constant isogeny, ϕ, there is a unique dual isogeny

$$\hat{\phi} : E_2 \longrightarrow E_1$$

such that $\hat{\phi} \circ \phi$ is equal to multiplication by n, where $n = \deg(\phi)$, on E_1 and $\phi \circ \hat{\phi}$ is multiplication by n on E_2. The existence of the dual isogeny implies that being isogenous is an equivalence relation on the set of all elliptic curves. We then have

LEMMA III.12 (Lemma 15.1 of [**25**]). *Two isogenous abelian varieties (and hence two isogenous elliptic curves) over a finite field have the same number of rational points.*

LEMMA III.13 (Lemma 8.4 of [**25**]). *Suppose $\phi : E \longrightarrow E'$ is a separable isogeny defined over K, whose kernel has exponent d, with d coprime to the characteristic of K. Assume that the elements of the kernel of ϕ and all the dth roots of unity are defined over K. Then all the elements in $E'[\hat{\phi}]$ are also defined over K and there is a natural non-degenerate pairing*

$$e_\phi : E[\phi] \times E'[\hat{\phi}] \longrightarrow \mu_{\mathbf{d}}(K).$$

When the isogeny in the previous lemma is equal to the multiplication-by-m map the pairing is the Weil pairing mentioned earlier. The above pairing is sometimes referred to as the ϕ-Weil pairing.

The set of all isogenies from a curve to itself, together with the zero map, form a ring. This is the *ring of endomorphisms* of E, denoted by $\mathrm{End}(E)$. Clearly $\mathrm{End}(E)$ contains a subring isomorphic to $\mathbb{Z}$, as multiplication by m is an isogeny from E to E of degree m^2. There are three possibilities for the structure of the ring $\mathrm{End}(E)$ (see [**147**, Section III.9]).

1. $\mathrm{End}(E) = \mathbb{Z}$; this does not occur for elliptic curves over a finite field.

2. $\mathrm{End}(E)$ is an order in an imaginary quadratic field. Over finite fields such curves are called ordinary.
3. $\mathrm{End}(E)$ is the maximal order in a quarternion algebra. Over finite fields such curves are called supersingular but over fields of characteristic zero this case does not occur.

Recall that a curve, E, is supersingular over a field, $\mathbb{F}_q$, of characteristic p if and only if

- $p = 2$ or 3 and $j(E) = 0$.
- $p \geq 5$ and the trace of Frobenius satisfies $t = 0$.

In all characteristics we have that E is supersingular if and only if p divides the trace of Frobenius. If the curve has an endomorphism ring which is strictly larger than $\mathbb{Z}$, then the curve is said to possess *complex multiplication* (CM).

Now let l be a prime different from the characteristic of K and consider the l-power torsion, $E[l^n]$, for some fixed value of n. The group $E[l^n]$ can clearly be considered as a $(\mathbb{Z}/l^n\mathbb{Z})$-module of rank two. The absolute Galois group, $G = \mathrm{Gal}(\overline{K}/K)$, acts on $E[l^n]$ as a linear map. So we obtain a Galois representation:

$$\rho_{l,n} : G \longrightarrow \mathrm{Aut}(E[l^n]) \subset GL_2(\mathbb{Z}/l^n\mathbb{Z}).$$

We can also consider all l-power torsion at once by taking the Tate module (see [**147**])

$$T_l(E) = \lim_{\leftarrow} E[l^n].$$

This is a rank two $\mathbb{Z}_l$-module, where $\mathbb{Z}_l$ is the l-adic integers. The inverse limit used to produce T_l is 'compatible' with the inverse limit used to define the absolute Galois group G, in the sense that $\rho_{l,n}$ will factor through a finite quotient group of G. Hence, we obtain a continuous l-adic Galois representation:

$$\rho_l : G \longrightarrow \mathrm{Aut}(T_l(E)) \subset GL_2(\mathbb{Z}_l).$$

If $K = \mathbb{Q}$ then sitting inside G are special elements, for each prime p, called the Frobenius elements. These are defined up to conjugation and their images generate the quotient of their decomposition group by the inertia group, $D_p/I_p = \mathrm{Gal}(\overline{\mathbb{F}_p}/\mathbb{F}_p)$. We then look at the image under ρ_l of a Frobenius element, σ_p, if the curve is non-singular over $\mathbb{F}_p$. The element $\rho_l(\sigma_p)$ is a matrix whose characteristic polynomial is well defined and independent of l. The trace of $\rho(\sigma_p)$ we denote by t_p and is the trace of Frobenius at the prime p.

If $K = \mathbb{F}_q$ then G is generated by the Frobenius element σ_q. The element $\rho_l(\sigma_p)$ is also a matrix whose characteristic polynomial is well defined and independent of l. Its trace is the trace of Frobenius, t, mentioned earlier.

III.7. Various Functions and q-Expansions

It is a standard fact [**147**], used in complex analysis, the theory of partial differential equations and number theory, that an elliptic curve over $\mathbb{C}$ defines a lattice in $\mathbb{C}$ (and hence a torus). The lattice will be denoted by $\Lambda = \mathbb{Z}\omega_1 + \mathbb{Z}\omega_2$, where $\omega_1, \omega_2 \in \mathbb{C}$ are the periods of the associated, doubly periodic *Weierstrass $\wp$-function*

$$\wp(z) = \frac{1}{z^2} + \sum_{\omega \in \Lambda \setminus 0} \left(\frac{1}{(z-\omega)^2} - \frac{1}{\omega^2} \right).$$

This function satisfies the differential Equation (III.1).

The periods, ω_1 and ω_2, can be suitably chosen so that the quantity

$$\tau = \frac{\omega_1}{\omega_2}$$

lies in the upper half of the complex plane, $\mathcal{H} = \{z \in \mathbb{C} : \mathrm{Im}(z) > 0\}$. The map from $\mathbb{C}$ (modulo Λ) to points on the corresponding elliptic curve is given by

$$\begin{array}{rcl} \mathbb{C}/\Lambda & \longrightarrow & E \\ z + \Lambda & \longmapsto & \begin{cases} (\, x_\Lambda,\ (\wp'(z) - a_1 x_\Lambda - a_3)/2\,), & z \notin \Lambda, \\ \mathcal{O}, & z \in \Lambda. \end{cases} \end{array}$$

where $x_\Lambda = \wp(z) - b_2/12$. The codomain of this map corresponds to the long Weierstrass form of the curve. The special case

$$z + \Lambda \mapsto (\wp(z), \wp'/2), \quad z \notin \Lambda,$$

corresponds to the short form $Y^3 = X^3 + aX + b$. The coefficients of the short form are obtained with the formulae

$$g_2 = 60 \sum_{\omega \in \Lambda \setminus \{0\}} \frac{1}{\omega^4}, \qquad g_3 = 140 \sum_{\omega \in \Lambda \setminus \{0\}} \frac{1}{\omega^6},$$

and $a = -g_2/\sqrt[3]{4}$, $b = -g_3$. The inverse correspondence, leading from the coefficients of the curve to the periods ω_1 and ω_2, can also be computed (see, for instance, [**29**]).

The complex number $\tau \in \mathcal{F}$ characterizes elliptic curves up to isomorphism, i.e. if $\tau = \omega_1/\omega_2 = \omega_1'/\omega_2'$, then the elliptic curves derived from the lattices $\Lambda = \mathbb{Z}\omega_1 + \mathbb{Z}\omega_2$ and $\Lambda' = \mathbb{Z}\omega_1' + \mathbb{Z}\omega_2'$ are isomorphic. An elliptic curve over $\mathbb{C}$ associated to τ in this way is denoted by E_τ. We can also consider the j-invariant of the curve as a function on $\mathcal{H}$ and write

$$j(\tau) = j(E_\tau),$$

which is well defined due to the invariance of $j(E_\tau)$ under curve isomorphisms. What makes this function $j(\tau)$ so exciting is that it is one of the simplest examples of a modular function [**147**].

LEMMA III.14. *For any matrix*

$$A = \begin{pmatrix} a & b \\ c & d \end{pmatrix} \in SL_2(\mathbb{Z})$$

we have

$$j\left(\frac{a\tau+b}{c\tau+d}\right)=j(\tau).$$

Also, $j(\tau)$ is periodic of period one, and has the Fourier series

$$j(\tau)=\frac{1}{q}+744+\sum_{n\geq 1}c_nq^n,$$

where $q=e^{2\pi i\tau}$, and the c_n are positive integers.

Here, $SL_2(\mathbb{Z})$ is the special linear group of 2×2 matrices over the integers, of determinant 1. Any complex number τ^* is equivalent to a τ, under $SL_2(\mathbb{Z})$ transformations, which lies in the standard fundamental region for such transformations,

$$\mathcal{F}=\{\tau\in\mathbb{C}:\mathrm{Im}(\tau)>0,\ -1/2\leq \mathrm{Re}(\tau)<1/2,\ |\tau|\geq 1\}.$$

Therefore, by Lemma III.14, when considering E_τ, we can assume that τ is in $\mathcal{F}$.

We now present various functions and series which are defined via expansions in the variable $q=e^{2\pi i\tau}$ and are related to the j-invariant above. We shall use these functions in various places in the book, so it is convenient to have them defined in a single place. For example, we can define

$$\Delta(\tau)=q\prod_{n=1}^{\infty}(1-q^n)^{24},$$

where, again, $q=e^{2\pi i\tau}$. It can be shown that this series may be written as

$$\begin{aligned}\Delta(\tau)&=q\left(1+\sum_{n\geq 1}(-1)^n\left(q^{n(3n-1)/2}+q^{n(3n+1)/2}\right)\right)^{24},\\&=\sum_{n\geq 1}\tau_nq^n.\end{aligned}\qquad\text{(III.10)}$$

Also, as expected, the power series satisfies $\Delta(\tau)=\Delta(E_\tau)$, where the latter is the discriminant of the curve defined earlier in the chapter. The function $\Delta(\tau)$ is also related to $j(\tau)$ using the formulae

$$h(\tau)=\frac{\Delta(2\tau)}{\Delta(\tau)},\ j(\tau)=\frac{(256h(\tau)+1)^3}{h(\tau)}.$$

The coefficients τ_n of $\Delta(\tau)$ in Equation (III.10) define a function, $n\mapsto\tau_n$, called the Ramanujan τ-function. This is a very interesting number-theoretic function which has the following properties:

THEOREM III.15. *The following all hold for the function τ_n:*

- *It is multiplicative, in the sense that if m and n are coprime then*

$$\tau_{mn}=\tau_m\tau_n.$$

- *If p is a prime and $t \geq 1$ then*

$$\tau_{p^{t+1}} = \tau_p \tau_{p^t} - p^{11} \tau_{p^{t-1}}.$$

- *For all $n \geq 1$*

$$|\tau_n| \leq \sigma_0(n) n^{11/2}$$

where $\sigma_0(n)$ denotes the number of positive divisors of n.

All of these results were conjectured by Ramanujan, the first two were proved by Mordell while the last was proved by Deligne. The function $\Delta(\tau)$ is itself the 24th power of a function of great importance, namely Dedekind's η-function

$$\begin{aligned} \eta(\tau) &= \Delta(\tau)^{1/24} = q^{1/24} \prod_{n=1}^{\infty} (1 - q^n) \\ &= e^{2\pi i \tau/24} \left(1 + \sum_{n \geq 1} (-1)^n \left(q^{n(3n-1)/2} + q^{n(3n+1)/2} \right) \right). \end{aligned}$$

The Dedekind η-function satisfies the following identities:

$$\eta(\tau + 1) = e^{2\pi i/24} \eta(\tau), \ \eta(-1/\tau) = \sqrt{-i\tau}\, \eta(\tau)$$

where the branch in the complex square root function is taken to be on the positive real axis. We will also require the following Eisenstein series, for $k = 0, 1, 2, \ldots$:

$$E_{2k}(\tau) = 1 - \frac{4k}{B_{2k}} \sum_{n \geq 1} \sigma_{2k-1}(n) q^n,$$

where B_i represents the ith Bernoulli number and $\sigma_i(n) = \sum_{d|n} d^i$. For example we have

$$\begin{aligned} E_2(\tau) &= 1 - 24 \sum_{n=1}^{\infty} \frac{n q^n}{1 - q^n}, \\ E_4(\tau) &= 1 + 240 \sum_{n=1}^{\infty} \frac{n^3 q^n}{1 - q^n}, \\ E_6(\tau) &= 1 - 504 \sum_{n=1}^{\infty} \frac{n^5 q^n}{1 - q^n}. \end{aligned}$$

These are related to $\Delta(\tau)$ by Jacobi's formula

$$\Delta(\tau) = \frac{E_4(\tau)^3 - E_6(\tau)^2}{1728}$$

and to the function $j(\tau)$ by

$$j(\tau) = \frac{E_4(\tau)^3}{\Delta(\tau)}.$$

III.8. Modular Polynomials and Variants

Modular polynomials play a significant role in the improvements by Atkin and Elkies to Schoof's point counting algorithm considered in Chapter VII, as well as in other more recent variants. The properties of these polynomials are reviewed here (without proof) drawing from the references [**148**], [**142**] and [**85**].

The correspondence between lattices $\mathbb{Z}\omega_1 + \mathbb{Z}\omega_2$, $\omega_1, \omega_2 \in \mathbb{C}$, and elliptic curves over $\mathbb{C}$ was noted in the previous section, as was the invariance of $j(\tau)$ under transformations of the form $\tau' = (a\tau + b)/(c\tau + d)$, where

$$\begin{pmatrix} a & b \\ c & d \end{pmatrix} \in SL_2(\mathbb{Z}).$$

More generally, for a matrix

$$\alpha = \begin{pmatrix} a & b \\ c & d \end{pmatrix} \in GL_2(\mathbb{R}), \ \det(\alpha) > 0,$$

define

$$j \circ \alpha(\tau) = j\left(\frac{a\tau + b}{c\tau + d}\right).$$

This is the j-invariant of the elliptic curve $\mathbb{C}/(\mathbb{Z} + \mathbb{Z}\tau')$ with $\tau' = (a\tau + b)/(c\tau + d)$.

For a positive integer n, define

$$D_n^* = \left\{ \begin{pmatrix} a & b \\ c & d \end{pmatrix} : a, b, c, d \in \mathbb{Z}, \ ad - bc = n, \ \gcd(a, b, c, d) = 1 \right\},$$

and

$$S_n^* = \left\{ \begin{pmatrix} a & b \\ 0 & d \end{pmatrix} \in D_n^* : d > 0, \ 0 \le b < d \right\}.$$

It can be shown that

$$\#S_n^* = n \prod_{p|n} (1 + \frac{1}{p})$$

where the product is over primes dividing n.

Notice that when $n = \ell$, a prime, we have $\#S_\ell^* = \ell + 1$. This case will be of special interest in the study of isogenies, and their application in the context of the point counting algorithms described in Chapter VII. The following lemma establishes a connection between the matrices S_n^*, and the j-invariants of images of isogenies of degree n from a given curve. It is adapted from a problem in [**148**].

LEMMA III.16. *Let E_1 and E_2 be two elliptic curves over $\mathbb{C}$, with j-invariants $j(E_1) = j(\tau)$ and $j(E_2)$, respectively, and let n be a positive integer. Then,*

$$j(E_2) = j \circ \alpha(\tau), \qquad \alpha \in S_n^*,$$

if and only if there is an isogeny from E_1 to E_2 whose kernel is cyclic of degree n.

Define the modular polynomial of order n, by the equation

$$\Phi_n(x, j) = \prod_{\alpha \in S_n^*} (x - j \circ \alpha).$$

It can be shown that $\Phi_n \in \mathbb{Z}[j][x]$ and, as a polynomial in two variables, $\Phi_n(x, y)$, it is symmetric and of degree $\# S_n^*$ in each variable. Notice that j in this equation is a formal function of τ, defined by its q-expansion. The previous lemma then implies that there is an isogeny of degree n, from E_1 to E_2, if and only if $\Phi_n(j(E_1), j(E_2)) = 0$.

In the case that $n = \ell$, a prime, there are precisely $\ell + 1$ subgroups of the group of ℓ-torsion points, $E[\ell]$ of a curve E. Each such subgroup is the kernel of an isogeny of degree ℓ, corresponding to one of the $\ell + 1$ matrices in S_ℓ^*. Equivalently, each such subgroup corresponds to an isogenous curve with a j-invariant which is a zero of the polynomial $\Phi_\ell(x, j)$.

It can be shown that the modular polynomial $\Phi_\ell(x, y)$ is equal to

$$x^{\ell+1} - x^\ell y^\ell + y^{\ell+1}$$

plus terms of the form $a_{ij} x^i y^j$, $i, j \leq \ell$, $i + j < 2\ell$, $a_{ij} \in \mathbb{Z}$. By the Kronecker congruence relation (see [**142**], [**148**] and [**60**]), we have

$$\Phi_\ell(x, y) \equiv (x^\ell - y)(x - y^\ell) \pmod{\ell}.$$

Note that while the degree of the modular polynomials $\Phi_\ell(x, j)$ is $\ell + 1$ in either variable (rather than the $(\ell^2 - 1)/2$ of the division polynomials), their integer coefficients can become very large as ℓ increases. For example, the modular polynomials for $\ell = 3$ and $\ell = 5$ are given by [**51**]:

$$\begin{aligned}
\Phi_3(x, y) = \ & x^4 - x^3 y^3 + y^4 + 2232\,(x^3 y^2 + y^3 x^2) - 1069956\,(x^3 y + y^3 x) \\
& + 36864000\,(x^3 + y^3) + 2587918086\,x^2 y^2 \\
& + 8900222976000\,(x^2 y + y^2 x) \\
& + 452984832000000\,(x^2 + y^2) - 770845966336000000\,xy \\
& + 1855425871872000000000\,(x + y),
\end{aligned}$$

$$
\begin{aligned}
\Phi_5(x,y) = \; & x^6 - x^5y^5 + y^6 \\
& +3720\,(x^5y^4 + x^4y^5) \\
& -4550940\,(x^5y^3 + x^3y^5) \\
& +2028551200\,(x^5y^2 + x^2y^5) \\
& -246683410950\,(x^5y + xy^5) \\
& +1963211489280\,(x^5 + y^5) \\
& +1665999364600\,x^4y^4 \\
& +107878928185336800\,(x^4y^3 + x^3y^4) \\
& +383083609779811215375\,(x^4y^2 + x^2y^4) \\
& +128541798906828816384000\,(x^4y + xy^4) \\
& +1284733132841424456253440\,(x^4 + y^4) \\
& -441206965512914835246100\,x^3y^3 \\
& +26898488858380731577417728000\,(x^3y^2 + x^2y^3) \\
& -192457934618928299655108231168000\,(x^3y + xy^3) \\
& +280244777828439527804321565297868800\,(x^3 + y^3) \\
& +5110941777552418083110765199360000\,x^2y^2 \\
& +36554736583949629295706472332656640000\,(x^2y + xy^2) \\
& +6692500042627997708487149415015068467200\,(x^2 + y^2) \\
& -264073457076620596259715790247978782949376\,xy \\
& +53274330803424425450420160273356509151232000\,(x + y) \\
& +141359947154721358697753474691071362751004672000\,.
\end{aligned}
$$

The rate of growth of the coefficients of Φ_ℓ was characterized by Cohen in [**31**], after initial estimates by Mahler ([**91**] [**92**]). Let $h(\Phi_\ell)$ denote the natural logarithm of the largest absolute value of a coefficient of $\Phi_\ell(x,y)$ (for example, for Φ_5 above, we have $h(\Phi_5)\approx 108.6$, attained by its constant coefficient). When ℓ is prime, it follows from the results in [**31**] that

$$h(\Phi_\ell) = 6(\ell+1)\left((1-\frac{2}{\ell})\log\ell + O(1)\right).$$

In our applications to point counting algorithms for elliptic curves over $\mathbb{F}_q$, ℓ will be a prime taking on values of the order of $\log q$, with q being a prime (or a power of two) with binary expansion a few hundreds of bits long. Although the coefficients of the modular polynomials are eventually reduced modulo the characteristic of the field, they are often computed first over $\mathbb{Z}$. Assuming, for instance, that the binary expansion of q is about two hundred bits long, the bound above places the coefficients of Φ_ℓ at about 30 times the binary length of q, a heavy computational burden indeed.

To overcome difficulties posed by the large coefficients some authors have given alternative modular polynomials. However, even these need to be computed with care. We give one such variant below. Other variants are described, for example, in [**108**] and [**40**].

Let s be the least positive integer such that $v = s(\ell-1)/12 \in \mathbb{Z}_{>0}$. Hence, $s = 12/\gcd(\ell-1, 12)$. Define

$$f(\tau) = \left(\frac{\eta(\tau)}{\eta(\ell\tau)}\right)^{2s},$$

where $\eta(z)$ is Dedekind's η-function. We then have the following theorem, which allows us to define variants of the modular polynomials which are more suited to computations.

THEOREM III.17 (see [**110**]). *There exist coefficients $a_{r,k} \in \mathbb{Z}$ such that*

$$\sum_{r=0}^{\ell+1}\sum_{k=0}^{v} a_{r,k} \cdot j(\ell\tau)^k \cdot f(\tau)^r = 0.$$

Define the polynomial

$$G_\ell(x,y) = \sum_{r=0}^{\ell+1}\sum_{k=0}^{v} a_{r,k}x^r y^k \in \mathbb{Z}[x,y].$$

Let E be an elliptic curve defined over $\mathbb{F}_q$. Then, when interpreted over $\mathbb{F}_q$, the polynomial $G_\ell(x, j(E))$ has the same splitting type over $\mathbb{F}_q$ as the ℓth modular polynomial $\Phi_\ell(x, j(E))$.

In Theorem III.17, 'splitting type' refers to the degrees and multiplicities of the irreducible factors of the polynomials over $\mathbb{F}_q$.

It turns out that considerably less precision is required for constructing these polynomials since their coefficients are much smaller than those of the standard modular polynomials. This property means they are easier to compute and store than the polynomials $\Phi_\ell(x,y)$, and it also leads to significant (and crucial) savings in the performance of the algorithms that require their use. Detailed explanations of the computation of these alternative polynomials are given in [**110**] and [**74**]. We shall just summarize the method. The idea is to compute the functions $s_r(\tau)$, for $r = 0, \ldots, \ell+1$, given by

$$s_r(\tau) = \sum_{k=0}^{v} a_{\ell+1-r,k}\, j(\ell\tau)^k \tag{III.11}$$

since the polynomial given by

$$h(X) = \sum_{r=0}^{\ell+1} s_r(\tau) X^{\ell+1-r}$$

has the roots, for $n = 0, \ldots, \ell-1$,

$$f\left(\tau + \frac{n}{\ell}\right) \text{ and } \left(\frac{\ell^s}{f(\ell\tau)}\right).$$

We can compute $s_r(\tau)$ in the following way, keeping all functions of τ in terms of their q-expansions. First compute the coefficients, b_i, in

$$f(\tau)^r = q^{-vr}\left(\sum_{i=0}^{\infty} b_i q^i\right),$$

and then

$$\begin{aligned} c_{r,1}(\tau) &= \sum_{n=0}^{\ell-1} f\left(\tau + \frac{n}{\ell}\right)^r \\ &= q^{-vr}\sum_{i=0}^{\infty} \ell_i b_i q^i. \end{aligned}$$

where $\ell_i = \ell$ if $i \equiv vr \pmod{\ell}$ and $\ell_i = 0$ otherwise. We can then compute the sum of the rth powers of the roots of $h(X)$ using the formula

$$c_r(\tau) = c_{r,1}(\tau) + \left(\frac{\ell^s}{f(\ell\tau)}\right)^r.$$

Then using Newton's formulae we can express the $s_r(\tau)$ using the iteration, for $i = 1, \dots, \ell+1$,

$$s_r(\tau) = -\sum_{i=0}^{r} c_{r-i}(\tau) s_i(\tau),$$

where $s_0(\tau) = -1$. We can then compute the desired coefficients $a_{\ell+1-r,k}$ from Equation (III.11), for $r = 1, \dots, \ell+1$ and $k = 0, \dots v$. For the other coefficients we have $a_{\ell+1,k} = 0$ for $k = 1, \dots, v$ and $a_{\ell+1,0} = 1$.

With the coefficients $a_{r,k}$ known, the bivariate polynomial $G_\ell(x, y)$ can be found from the equation given in Theorem III.17 by replacing $j(\ell\tau)$ with y.

As an example of how much smaller the coefficients of these polynomials are, compared to the standard modular polynomials, consider the examples

$$\begin{aligned} G_3(x,y) &= 729 + 756x - xy + 270x^2 + 36x^3 + x^4, \\ G_5(x,y) &= 125 + 750x - xy + 1575x^2 + 1300x^3 + 315x^4 + 30x^5 + x^6. \end{aligned}$$

However, although the polynomials are computed over $\mathbb{C}$, if we are working over fields of characteristic two then we may only need to reduce the integer coefficients of the modular polynomials modulo two once and for all and store them in a table. So for fields of characteristic two we will still use the standard modular polynomials. For example, on computing the first few modular polynomials, we find that modulo two they are given by

$$\begin{aligned} \Phi_3(x,y) &\equiv x^4 + x^3y^3 + y^4 \pmod{2}, \\ \Phi_5(x,y) &\equiv x^6 + x^5y^5 + x^4y^2 + x^2y^4 + y^6 \pmod{2}, \\ \Phi_7(x,y) &\equiv x^8 + x^7y^7 + x^6y^6 + y^8 \pmod{2}, \\ \Phi_{11}(x,y) &\equiv x^{12} + x^{11}y^{11} + x^{11}y^3 + x^{10}y^6 + x^9y^9 + x^8y^4 + x^6y^{10} \\ &\quad + x^4y^8 + x^3y^{11} + y^{12} \pmod{2}, \end{aligned}$$

$$\begin{aligned}\Phi_{13}(x,y) &\equiv x^{14}+x^{13}y^{13}+x^{13}y^{5}+x^{12}y^{2}+x^{10}y^{4}+x^{8}y^{6}+x^{6}y^{8}+x^{5}y^{13}\\ &\quad +x^{4}y^{10}+x^{2}y^{12}+y^{14} \pmod 2,\\ \Phi_{17}(x,y) &\equiv x^{18}+x^{17}y^{17}+x^{17}y^{9}+x^{16}y^{2}+x^{16}y^{10}+x^{14}y^{12}+x^{12}y^{14}\\ &\quad +x^{10}y^{16}+x^{9}y^{17}+x^{2}y^{16}+y^{18} \pmod 2.\end{aligned}$$

CHAPTER IV

Efficient Implementation of Elliptic Curves

The basic building blocks of an elliptic curve cryptosystem over $\mathbb{F}_q$ are computations of the form

$$Q = [k]P = \underbrace{P + P + \cdots + P}_{k \text{ times}}, \tag{IV.1}$$

where P is a curve point, and k is an arbitrary integer in the range $1 \leq k < \text{ord}(P)$. For some of the cryptographic protocols, P is a designated fixed point that generates a large, prime order subgroup of $E(\mathbb{F}_q)$, while for others P is an arbitrary point in such a subgroup. The strength of the cryptosystem lies in the fact that given the curve, the point P (be it fixed or arbitrary) and $[k]P$, it is hard to recover k. This is the elliptic curve discrete logarithm problem (ECDLP), which is discussed at length in Chapter V.

We refer to the computation of Equation (IV.1) as *point multiplication.* Efficient algorithms for this computation are the subject of this chapter. We start by analysing the computational complexity of the group operation.

IV.1. Point Addition

As noted in Chapter III, the simplified formulae for the group law take on different forms depending on the characteristic of the underlying field. We analyse the computational complexity of these formulae separately for characteristic $p > 3$, and for characteristic two.

IV.1.1. Fields of characteristic $p > 3$. Affine coordinates. We recall from Chapter III the formulae for point addition on a curve

$$E : Y^2 = X^3 + aX + b$$

with $a, b \in \mathbb{F}_q$, $q = p^n$, p a prime greater than three. Let $P_1 = (x_1, y_1)$ and $P_2 = (x_2, y_2)$ be points in $E(\mathbb{F}_q)$ given in affine coordinates, and where some convention is used to represent $\mathcal{O}$. Assume $P_1, P_2 \neq \mathcal{O}$, and $P_1 \neq -P_2$, conditions that are all easily checked. The sum $P_3 = (x_3, y_3) = P_1 + P_2$ can be computed as follows.

If $P_1 \neq P_2$,

$$\begin{aligned} \lambda &= \frac{y_2 - y_1}{x_2 - x_1}, \\ x_3 &= \lambda^2 - x_1 - x_2, \\ y_3 &= (x_1 - x_3)\lambda - y_1. \end{aligned}$$

If $P_1 = P_2$,

$$\begin{aligned} \lambda &= \frac{3x_1^2 + a}{2y_1}, \\ x_3 &= \lambda^2 - 2x_1, \\ y_3 &= (x_1 - x_3)\lambda - y_1. \end{aligned}$$

When $P_1 \neq P_2$, the computation requires one field inversion and three field multiplications. We will denote this computational cost by $1I + 3M$, where I and M denote, respectively, the cost of field inversion and multiplication. Squarings are counted as regular multiplications. When $P_1 = P_2$, the cost of the point doubling is $I + 4M$. We neglect the cost of field additions, as well as the cost of multiplication by small constants (e.g., 2 and 3 in the computation of λ when $P_1 = P_2$).

Projective coordinates. In cases where field inversions are significantly more expensive than multiplications, it is efficient to implement *projective* coordinates. The conventional projective (or *homogeneous*) coordinates were introduced in Chapter III. A projective point (X, Y, Z) on the curve satisfies the homogeneous Weierstrass equation

$$Y^2 Z = X^3 + aXZ^2 + bZ^3,$$

and, when $Z \neq 0$, it corresponds to the affine point $(X/Z, Y/Z)$. It turns out that other projective representations lead to more efficient implementations of the group operation [**27**]. In particular, we will prefer a *weighted projective* representation (also referred to as *Jacobian* representation – [**27**] [**30**]), where a triplet (X, Y, Z) corresponds to the affine coordinates $(X/Z^2, Y/Z^3)$ whenever $Z \neq 0$. This is equivalent to using a weighted projective curve equation of the form

$$Y^2 = X^3 + aXZ^4 + bZ^6.$$

The point at infinity $\mathcal{O}$ is represented by any triplet $(\gamma^2, \gamma^3, 0)$, $\gamma \in \mathbb{F}_q^*$, although in a practical implementation, since the coordinates of this point are never actually operated on, any triplet with $Z = 0$ would do. Weighted projective coordinates are very natural for elliptic curves. For example, for the division polynomial sequences $\psi_m(x, y), \theta_m(x, y), \omega_m(x, y)$ defined in Section III.4, we have $[m](X, Y, Z) = (\theta_m(X, Y), \omega_m(X, Y), \psi_m(X, Y))$. For the remainder of the chapter, and for the sake of conciseness, we will slightly abuse terminology and use the term 'projective' to mean 'weighted projective'. Conversion from affine to projective coordinates is trivial, while conversion in the other direction costs $1I + 4M$.

The key observation is that point addition can be done in projective coordinates using field multiplications only, with no inversions required. Thus, inversions are deferred, and only one need be performed at the end of a point multiplication operation, if it is required that the final result be given in affine coordinates. The cost of eliminating inversions is an increased number of multiplications, so the appropriateness of using projective coordinates is strongly determined by the ratio $I : M$.

The computation sequences in Figures IV.1 and IV.2 are adapted from the description in the appendices to the IEEE P1363 draft standard, [**P1363**]. A discussion of these sequences, together with similar ones for conventional homogeneous coordinates, and a comparison between the two types of coordinates can be found in [**27**]. This reference, as well as [**30**], also discusses various (redundant) mixed representations, e.g. (X, Y, Z, Z^2, Z^3), which may have some computational advantages.

The sequence in Figure IV.1 computes the sum $P_3 = (X_3, Y_3, Z_3)$ of two points $P_1 = (X_1, Y_1, Z_1)$ and $P_2 = (X_2, Y_2, Z_2)$ in projective coordinates. We assume that $P_1, P_2 \neq \mathcal{O}$, and that $P_1 \neq \pm P_2$. The latter condition is easily checked at an early stage of the computation, as discussed below. In the figure, the cost of each step of the computation is noted at the right-hand side of the step.

FIGURE IV.1. Point addition in projective coordinates, characteristic $p > 3$.

$$
\begin{array}{lcll}
\lambda_1 & = & X_1 Z_2^2 & 2M \\
\lambda_2 & = & X_2 Z_1^2 & 2M \\
\lambda_3 & = & \lambda_1 - \lambda_2 & \\
\lambda_4 & = & Y_1 Z_2^3 & 2M \\
\lambda_5 & = & Y_2 Z_1^3 & 2M \\
\lambda_6 & = & \lambda_4 - \lambda_5 & \\
\lambda_7 & = & \lambda_1 + \lambda_2 & \\
\lambda_8 & = & \lambda_4 + \lambda_5 & \\
Z_3 & = & Z_1 Z_2 \lambda_3 & 2M \\
X_3 & = & \lambda_6^2 - \lambda_7 \lambda_3^2 & 3M \\
\lambda_9 & = & \lambda_7 \lambda_3^2 - 2X_3 & \\
Y_3 & = & (\lambda_9 \lambda_6 - \lambda_8 \lambda_3^3)/2 & \underline{3M} \\
 & & & 16M
\end{array}
$$

The total cost for general point addition is $16M$. A special case of interest arises when $Z_1 = 1$, i.e., one point is given in affine coordinates, and the other one in projective coordinates. This case, which will occur in the point multiplication algorithms, costs $11M$, and will be referred to as a *mixed* addition.

The condition $P_1 = \pm P_2$ is equivalent to $\lambda_3 = 0$ in Figure IV.1. Furthermore, given that $\lambda_3 = 0$, the condition $P_1 = P_2$ is equivalent to $\lambda_6 = 0$. When this condition is detected, a point doubling routine is used, shown in Figure IV.2. The point doubling computation costs $10M$. This can be reduced to $8M$ when $a = -3$, as in this case the computation of λ_1 can be

FIGURE IV.2. Point doubling in projective coordinates, characteristic $p > 3$.

$$\begin{array}{rcll}
\lambda_1 &=& 3X_1^2 + aZ_1^4 & 4M \\
Z_3 &=& 2Y_1Z_1 & 1M \\
\lambda_2 &=& 4X_1Y_1^2 & 2M \\
X_3 &=& \lambda_1^2 - 2\lambda_2 & 1M \\
\lambda_3 &=& 8Y_1^4 & 1M \\
Y_3 &=& \lambda_1(\lambda_2 - X_3) - \lambda_3 & 1M \\
\hline
 & & & 10M
\end{array}$$

rearranged as $\lambda_1 = 3(X_1 - Z_1^2)(X_1 + Z_1^2)$, costing $2M$ instead of $4M$. By the characterization of isomorphisms in Section III.3.1, a curve $E_{a,b}$ can be transformed into an $\mathbb{F}_q$-isomorphic one $E_{a',b'}$ with $a' = -3$ if and only if $-3/a$ has a fourth root in $\mathbb{F}_q$. This holds for about a quarter of the values of a when $q \equiv 1 \pmod 4$, and one half of the values when $q \equiv 3 \pmod 4$.

The different costs for point addition and doubling in characteristic $p > 3$ are summarized in Table IV.1. We observe in the table that the cost of point doubling in projective coordinates is about a half of that of a general addition (when $a = -3$), whereas in affine coordinates doubling is the more expensive operation.

TABLE IV.1. Cost of point addition, characteristic $p > 3$.

Operation	Coordinates		
	affine	mixed	projective
General addition	$1I + 3M$	$11M$	$16M$
Doubling (arbitrary a)	$1I + 4M$	n/a	$10M$
Doubling ($a = -3$)	$1I + 4M$	n/a	$8M$

IV.1.2. Fields of characteristic two. Affine coordinates. Recall from Chapter III the formulae for point addition on a curve

$$E : Y^2 + XY = X^3 + a_2X^2 + a_6$$

with $a_2, a_6 \in \mathbb{F}_q$, $q = 2^n$, $a_6 \neq 0$. Let $P_1 = (x_1, y_1)$ and $P_2 = (x_2, y_2)$ be points in $E(\mathbb{F}_q)$ given in affine coordinates, where some convention is used to represent $\mathcal{O}$ (in this case, $(0, 0)$ can be used for that purpose since such a point is never on the curve). Assume $P_1, P_2 \neq \mathcal{O}$, and $P_1 \neq -P_2$. The sum $P_3 = (x_3, y_3) = P_1 + P_2$ is computed as follows.

If $P_1 \neq P_2$,

$$\begin{aligned}\lambda &= \frac{y_1 + y_2}{x_1 + x_2},\\ x_3 &= \lambda^2 + \lambda + x_1 + x_2 + a_2,\\ y_3 &= (x_1 + x_3)\lambda + x_3 + y_1.\end{aligned}$$

If $P_1 = P_2$,

$$\begin{aligned}\lambda &= \frac{y_1}{x_1} + x_1,\\ x_3 &= \lambda^2 + \lambda + a_2,\\ y_3 &= (x_1 + x_3)\lambda + x_3 + y_1.\end{aligned}$$

In either case, the computation requires one field inversion, two field multiplications, and one squaring, or $1I + 2M + 1S$. In the case of characteristic two, the cost of a squaring operation, denoted by S, is much lower than that of a general multiplication. Therefore, squarings are counted separately, and in fact, we will later on neglect their cost completely.

Projective coordinates. As in the case of characteristic $p > 3$, we will use weighted projective coordinates, where a projective point (X, Y, Z), $Z \neq 0$, maps to the affine point $(X/Z^2, Y/Z^3)$. This corresponds to using a weighted projective curve equation of the form

$$Y^2 + XYZ = X^3 + a_2X^2Z^2 + a_6Z^6.$$

Conversion from projective to affine coordinates costs, in this case, $1I + 3M + 1S$. The computation sequences for point addition in this representation are presented in Figures IV.3 and IV.4. They are adapted, as before, from [**P1363**].

The total cost for general point addition is $15M + 5S$. This is reduced to $14M + 4S$ when $a_2 = 0$, which accounts for one of the two isomorphism classes of non-supersingular elliptic curves over $\mathbb{F}_{2^n}$. The mixed-addition case where $Z_1 = 1$ costs, in the case of characteristic two, $11M + 4S$ ($10M + 3S$ when $a_2 = 0$).

As in the odd characteristic case, the condition $P_1 = \pm P_2$ is equivalent to $\lambda_3 = 0$, then $P_1 = P_2$ is equivalent to $\lambda_6 = 0$. The detection of the conditions $P_1 = \pm P_2$ is similar to the odd characteristic case. The point doubling routine is shown in Figure IV.4, where the field element d_6 is defined as $d_6 = \sqrt[4]{a_6} = a_6^{2^{n-2}}$. The point doubling computation costs $5M + 5S$. Notice that, since squaring is much faster than general multiplication in characteristic two, point doubling in projective coordinates is close to three times as fast as general point addition. This is contrasted with the affine case, where both operations are of about the same arithmetic complexity.

The different costs for point addition and doubling in characteristic two are summarized in Table IV.2.

FIGURE IV.3. Point addition in projective coordinates, characteristic 2.

$$\begin{array}{lcll}
\lambda_1 & = & X_1 Z_2^2 & 1M + 1S \\
\lambda_2 & = & X_2 Z_1^2 & 1M + 1S \\
\lambda_3 & = & \lambda_1 + \lambda_2 & \\
\lambda_4 & = & Y_1 Z_2^3 & 2M \\
\lambda_5 & = & Y_2 Z_1^3 & 2M \\
\lambda_6 & = & \lambda_4 + \lambda_5 & \\
\lambda_7 & = & Z_1 \lambda_3 & 1M \\
\lambda_8 & = & \lambda_6 X_2 + \lambda_7 Y_2 & 2M \\
Z_3 & = & \lambda_7 Z_2 & 1M \\
\lambda_9 & = & \lambda_6 + Z_3 & \\
X_3 & = & a_2 Z_3^2 + \lambda_6 \lambda_9 + \lambda_3^3 & 3M + 2S \\
Y_3 & = & \lambda_9 X_3 + \lambda_8 \lambda_7^2 & 2M + 1S \\
\hline
 & & & 15M + 5S
\end{array}$$

FIGURE IV.4. Point doubling in projective coordinates, characteristic 2.

$$\begin{array}{lcll}
Z_3 & = & X_1 Z_1^2 & 1M + 1S \\
X_3 & = & (X_1 + d_6 Z_1^2)^4 & 1M + 2S \\
\lambda & = & Z_3 + X_1^2 + Y_1 Z_1 & 1M + 1S \\
Y_3 & = & X_1^4 Z_3 + \lambda X_3 & 2M + 1S \\
\hline
 & & & 5M + 5S
\end{array}$$

TABLE IV.2. Cost of point addition, characteristic 2.

Operation	Coordinates		
	affine	mixed	projective
General addition ($a_2 \neq 0$)	$1I + 2M + 1S$	$11M + 4S$	$15M + 5S$
General addition ($a_2 = 0$)	$1I + 2M + 1S$	$10M + 3S$	$14M + 4S$
Doubling	$1I + 2M + 1S$	n/a	$5M + 5S$

IV.2. Point Multiplication

Point multiplication in elliptic curves is a special case of the general problem of exponentiation in abelian groups. As such, it benefits from all the techniques available for the general problem, and the related *shortest addition chain* problem for integers. The latter is defined as follows. Let k be a positive integer (the input). Starting from the integer 1, and computing at each step the sum of two previous results, what is the least number of steps required to reach k?

Efficient algorithms for group exponentiation have received much attention by researchers in recent years, owing to their central role in public key cryptography (see Chapter I). The interest in the problem, however, is ancient. An excellent technical and historical account of exponentiation and the addition chain problem is given by Knuth [**61**, Ch. 4], who traces the problem back to 200 BC. The survey by Gordon [**48**] describes various fast methods, including some specialized to elliptic curve groups. Various techniques and algorithms for exponentiation in the context of cryptography are described, in fairly compact but detailed algorithmic form, in [**99**].

Although general methods of exponentiation can be used to compute point multiplication, certain idiosyncrasies of the elliptic curve version of the problem can be taken into account to obtain faster algorithms. First, elliptic curve subtraction has virtually the same cost as addition, so the search space for fast algorithms can be expanded to include *addition–subtraction* chains and *signed representations*, which are discussed in Sections IV.2.4–IV.2.5. Second, in tuning-up algorithms, the relative complexities of general point addition and point doubling have to be considered. As we saw in Section IV.1, this relation depends on the coordinate system used, and on the relative complexities of field inversion and multiplication. Third, for certain families of elliptic curves, specific shortcuts are available that can significantly reduce the computational cost of point multiplication. An example of such a family and the associated shortcuts is discussed in Section IV.3.

For the sake of concreteness, when analysing computational complexity in the remainder of the section, we will focus on the case of finite fields of characteristic two. Also, for simplicity, we will neglect the cost of squarings in these fields. The main ideas and the analysis, however, carry to other finite fields with only minor adjustments.

IV.2.1. The binary method. The simplest (and oldest) efficient method for point multiplication relies on the binary expansion of k.

ALGORITHM IV.1: **Point Multiplication: Binary Method.**

INPUT: A point P, an ℓ-bit integer $k = \sum_{j=0}^{\ell-1} k_j 2^j$, $k_j \in \{0,1\}$.
OUTPUT: $Q = [k]P$.

1. $Q \leftarrow \mathcal{O}$.
2. For $j = \ell-1$ to 0 by -1 do:
3. $\quad Q \leftarrow [2]Q$,
4. $\quad$ If $k_j = 1$ then $Q \leftarrow Q + P$.
5. Return Q.

The binary method requires $\ell-1$ point doublings and $W-1$ point additions (operations involving $\mathcal{O}$ are not counted), where ℓ is the length and W the weight (number of ones) of the binary expansion of k. Assuming that on the average $W = \ell/2$, that typically $\ell \approx n$, and neglecting $O(1)$ terms, the average number of field operations is $1.5nI + 3nM$ in affine representation, or $10nM$ in projective representation. We assume that P is given initially in affine representation, so Step 4 above involves a mixed addition costing $10M$ (we also assume $a_2 = 0$).

IV.2.2. The m-ary method. This method uses the m-ary expansion of k, where $m = 2^r$ for some integer $r \geq 1$. The binary method is a special case corresponding to $r = 1$.

ALGORITHM IV.2: **Point Multiplication: m-ary Method.**

INPUT: A point P, an integer $k = \sum_{j=0}^{d-1} k_j m^j$, $k_j \in \{0, 1, \ldots, m-1\}$.
OUTPUT: $Q = [k]P$.

Precomputation.

1. $P_1 \leftarrow P$.
2. For $i = 2$ to $m-1$ do $P_i \leftarrow P_{i-1} + P$. (We have $P_i = [i]P$.)
3. $Q \leftarrow \mathcal{O}$.

Main loop.

4. For $j = d-1$ to 0 by -1 do:
5. $\quad Q \leftarrow [m]Q$. (This requires r doublings.)
6. $\quad Q \leftarrow Q + P_{k_j}$.
7. Return Q.

It can be readily verified that the algorithm computes $[k]P$, following Horner's rule [**61**]:

$$[m](\ldots [m]([m]([k_{\ell-1}]P) + [k_{\ell-2}]P) + \cdots) + [k_0]P = [k]P.$$

The number of doublings in the main loop of the m-ary method is $(d-1)r$ (the first iteration is not counted, as it starts with $Q = \mathcal{O}$). Since $d = \lceil \ell/r \rceil$, where ℓ is the length of the binary representation of k, the number of doublings in the m-ary method may be up to $r-1$ less than the $\ell-1$ required by the binary method. For typical parameters, this is a rather modest gain in doublings, the main gains over the binary method being in the number of general point additions. The doublings in the main loop, however, can be exploited to obtain additional savings: by splitting the computation of $[m]Q$ into two stages, we can skip the even multiples of P in the precomputation phase. This leads to an improvement on the m-ary method, shown below. For this modification, we assume $r > 1$, otherwise we revert to the original binary method.

ALGORITHM IV.3: Point Multiplication: Modified m-ary Method.

```
INPUT:  A point P, an integer k = Σ_{j=0}^{d-1} k_j m^j,  k_j ∈ {0,1,...,m−1}.
OUTPUT: Q = [k]P.
Precomputation.
1.  P_1 ← P,  P_2 ← [2]P.
2.  For i = 1 to (m−2)/2 do P_{2i+1} ← P_{2i−1} + P_2.
3.  Q ← O.
Main loop.
4.  For j = d−1 to 0 by −1 do:
5.      If k_j ≠ 0 then do:
6.          Let s_j, h_j be such that k_j = 2^{s_j} h_j, h_j odd.
7.          Q ← [2^{r−s_j}]Q,
8.          Q ← Q + P_{h_j}.
9.      Else s_j ← r.
10.     Q = [2^{s_j}]Q.
11. Return Q.
```

It is readily verified that the modified m-ary method requires one point doubling and $2^{r-1}-1$ point additions in the precomputation phase, and at most $n-1$ point doublings and $d-1$ point additions in the main loop (to simplify the analysis, we take a pessimistic view, and ignore the fact that about one mth of the digits are expected to be zero and require no additions). Ignoring integer constraints for the purpose of estimating complexity, and setting $d = n/r$, the total number of curve operations is estimated at

$$N(n,r) = n + \frac{n}{r} + 2^{r-1} - 2. \qquad \text{(IV.2)}$$

The value of r minimizing $N(n,r)$ satisfies $r = \log_2 n - (2 - o(1)) \log_2 \log_2 n$. Substituting in Equation (IV.2) yields

$$N(n,r) = n + (1 + o(1))\frac{n}{\log_2 n},$$

which is asymptotically optimal for a generic addition chain method, due to a lower bound by Erdös [**41**]. This optimization is appropriate in affine coordinates, where additions and doublings have similar costs.

A slightly different optimization is required if we use projective coordinates. One possibility is to precompute the points P_2 and P_{2i+1}, $1 \le i \le (m-2)/2$, in affine coordinates, and then run the main loop in projective coordinates, using mixed addition for the operation $Q \leftarrow Q + P_{h_j}$ in Algorithm IV.3. The total cost of the point multiplication for the modified m-ary method is then estimated at

$$2^{r-1}(2M + I) + 10(\frac{n}{r} - 1)M + 5(n-1)M,$$

which can be optimized with respect to r given the ratio $I : M$. A similar expression can be derived for the case where projective coordinates are used throughout.

IV.2.3. Window methods. The m-ary scheme can be regarded as a special case of a *window* method, where bits of the multiplier k are processed in blocks (windows) of length r. In the m-ary methods of the previous section, the windows are contiguous and in fixed bit positions. A closer scrutiny of the modified m-ary method in Algorithm IV.3 reveals an inefficiency, due to the fact that trailing zeros are dropped from k_j (to obtain h_j), but new bits are not appended at the higher end, which is still constrained by the fixed m-ary digit boundary. Thus, higher values of h_j are less likely, and the array of precomputed points P_{h_j} is underutilized. This inefficiency is remedied in the following method, which processes windows up to length r disregarding fixed digit boundaries, and skips runs of zeros between them. These runs are taken care of by point doublings, which as we have seen, need to be computed in any case. As before, we assume $r > 1$.

ALGORITHM IV.4: **Point Multiplication: Sliding Window Method.**

INPUT: A point P, an integer $k = \sum_{j=0}^{\ell-1} k_j 2^j$, $k_j \in \{0,1\}$.
OUTPUT: $Q = [k]P$.
Precomputation.
1. $P_1 \leftarrow P$, $P_2 \leftarrow [2]P$.
2. For $i = 1$ to $2^{r-1}-1$ do $P_{2i+1} \leftarrow P_{2i-1} + P_2$.
3. $j \leftarrow \ell-1$, $Q \leftarrow \mathcal{O}$.
Main loop.
4. While $j \geq 0$ do:
5. If $k_j = 0$ then $Q \leftarrow [2]Q$, $j \leftarrow j-1$.
6. Else do:
7. Let t be the least integer such that $j - t + 1 \leq r$ and $k_t = 1$,
8. $h_j \leftarrow (k_j k_{j-1} \cdots k_t)_2$,
9. $Q \leftarrow [2^{j-t+1}]Q + P_{h_j}$,
10. $j \leftarrow t - 1$.
11. Return Q.

Using sliding windows has an effect equivalent to using fixed windows one bit larger, but without increasing the precomputation cost. An intuitive explanation for this effect is that the 'white space' of zeros between two consecutive sliding windows has an expected length of one, when we assume that the bits of k are obtained by independent tosses of a fair coin. Therefore, the total number of windows processed (and consequently, the number of general point additions in the main loop) behaves like $\ell/(r+1)$, as opposed to ℓ/r for the m-ary method. This fact is formally proven in [**71**].

The computational cost of the sliding window method is estimated at

$$(n + \frac{n}{r+1} + 2^{r-1} - 2)(2M + I)$$

for affine coordinates, and

$$2^{r-1}(2M + I) + (5n + 10\frac{n}{r+1} - 15)M$$

for projective/mixed coordinates.

IV.2.4. Signed Digit representations. As mentioned, subtraction has virtually the same cost as addition in the elliptic curve group. For the canonical curve equations of interest, the group negative of a point (x, y) is $(x, x+y)$ in characteristic two, and $(x, -y)$ in odd characteristic. This leads naturally to point multiplication methods based on addition–subtraction chains, which may reduce the number of curve operations.

Consider integer representations of the form $k = \sum_{j=0}^{\ell} s_j 2^j$, where $s_j \in \{-1, 0, 1\}$. We call this a (binary) *signed digit* (SD) representation. Clearly, this system includes the binary representation, so all integers k, $0 \leq k \leq 2^{\ell+1}-1$, are included, along with their negatives. But there are $3^{\ell+1}$ possible combinations, so the representation is clearly redundant. For example, the integer 3 can be represented as $(011)_2$ or $(10\overline{1})_2$, where $\overline{1} = -1$. As it turns out, this redundancy can be traded off for a sparsity constraint that results in more efficient point multiplication algorithms. We say that an SD representation is *sparse* if it has no adjacent non-zero digits, i.e. $s_j s_{j+1} = 0$ for all $j \geq 0$. A sparse SD representation is also called a *non-adjacent form* (NAF).

Several proofs of the following result can be found in the literature, starting with Reitwiesner [**131**]; see also [**28**], [**87**, Ch. 10] and [**109**].

LEMMA IV.1. *Every integer k has a unique NAF. The NAF has the lowest weight among all SD representations of k, and it is at most one digit longer than the shortest SD representation of k.*

The following algorithm computes the NAF of a non-negative integer given in binary representation. The description here follows [**99**]; other precursors and variants can be found in [**131**], [**93**], [**6**], [**87**, Ch. 10] and [**56**] (where the algorithm accepts general SD inputs).

ALGORITHM IV.5: **Conversion to NAF.**

INPUT: An integer $k = \sum_{j=0}^{\ell-1} k_j 2^j$, $k_j \in \{0, 1\}$.
OUTPUT: NAF $k = \sum_{j=0}^{\ell} s_j 2^j$, $s_j \in \{-1, 0, 1\}$.

```
1. c_0 ← 0.
2. For j = 0 to ℓ do:
3.     c_{j+1} ← ⌊(k_j + k_{j+1} + c_j)/2⌋  (assume k_i = 0 for i ≥ ℓ),
4.     s_j ← k_j + c_j − 2c_{j+1}.
5. Return (s_ℓ s_{ℓ−1} ··· s_0).
```

NAFs usually have fewer non-zero digits than binary representations. Morain and Olivos show in [**109**] that the expected weight of an NAF of length ℓ is $\ell/3$. The result is also proved in [**6**], where it is extended to m-ary SD representations, which have an expected weight $(m-1)\ell/(m+1)$.

The adaptation of the binary method for point multiplication to NAFs is straightforward: a subtraction is performed in lieu of an addition whenever a negative digit s_j is processed. Assuming an average NAF weight of $n/3$, the computation cost is $\frac{4}{3}n(2M+I)$ for affine coordinates, and $\frac{25}{3}nM$ for projective coordinates.

Clearly, fixed window and sliding window methods can be implemented for NAFs. The maximum possible absolute value of a NAF window of size r is $W_r = \frac{1}{3}(2^{r+1}-1)$ for r odd, and $W_r = \frac{1}{3}(2^{r+1}-2)$ for r even, given by the binary combinations $(1010\ldots101)$ and $(1010\ldots010)$ respectively. In the precomputation step, we need to compute and store points of the form $[i]P$, for $i = 2$ and all odd values of i, $3 \le i \le W_r$ (it is easily verified that W_r has the same parity as r). Thus, the number of point operations in the precomputation step is $\frac{1}{3}(2^r - (-1)^r)$. To estimate the expected number of point additions in the main loop of an NAF sliding window scheme, we consider the binary sequence obtained by taking the absolute values of the digits in the NAF. It follows from the results of [**109**] and [**6**] that such a sequence can be modelled by a Markov chain with transition probabilities $P(0|0) = P(1|0) = \frac{1}{2}$, $P(0|1) = 1$, $P(1|1) = 0$ where $P(a|b)$ denotes the probability of observing a symbol a immediately following a symbol b (we assume, as before, that the original integer k is drawn with uniform probability). Elementary analysis [**42**] of this transition matrix yields the expected length of a run of zeros between windows, which is given by a function

$$\nu(r) = \frac{4}{3} - \frac{(-1)^r}{3 \cdot 2^{r-2}}. \tag{IV.3}$$

Therefore, the expected number of point operations in an NAF sliding window scheme is estimated at

$$n + \frac{n+1}{r+\nu(r)} + \frac{2^r - (-1)^r}{3} - 2. \tag{IV.4}$$

A similar scheme, which uses a non-sparse SD representation, is analysed in [**69**]. The scheme produces SD representations of lower expected weights, but requires more precomputation, yielding what appears to be a slightly inferior trade off.

IV.2.5. A signed m-ary sliding window method. A slightly better asymptotic trade off can be obtained by using a signed m-ary scheme that is a natural extension of the sliding window method of Section IV.2.3. Although we have found no reference to this specific scheme in the literature, a suggestion to combine m-ary and signed methods appears in the closing remarks of [**109**].

In this method, we use a *non-redundant* signed m-ary representation, i.e., our digit set is $B = \{-2^{r-1}+1, \dots, -1, 0, 1, \dots, 2^{r-1}\}$ with windows of size up to r. We decompose the positive multiplier k as

$$k = \sum_{i=0}^{d-1} b_i 2^{e_i}, \quad b_i \in B \setminus \{0\},\ e_i \in \mathbb{Z}_{\geq 0}, \tag{IV.5}$$

where

$$e_{i+1} - e_i \geq r, \quad 0 \leq i \leq d-2. \tag{IV.6}$$

Such a decomposition is obtained by the following algorithm, which operates on the binary representation of k.

ALGORITHM IV.6: **Signed m-ary Window Decomposition**

INPUT: An integer $k = \sum_{j=0}^{\ell} k_j 2^j$, $k_j \in \{0,1\}$, $k_\ell = 0$.
OUTPUT: A sequence of pairs $\{(b_i, e_i)\}_{i=0}^{d-1}$.

1. $d \leftarrow 0$, $j \leftarrow 0$.
2. While $j \leq \ell$ do:
3. If $k_j = 0$ then $j \leftarrow j+1$.
4. Else do:
5. $t \leftarrow \min\{\ell, j+r-1\}$, $h_d \leftarrow (k_t k_{t-1} \cdots k_j)_2$.
6. If $h_d > 2^{r-1}$ then do:
7. $b_d \leftarrow h_d - 2^r$,
8. increment the number $(k_\ell k_{\ell-1} \cdots k_{t+1})_2$ by 1.
9. Else $b_d \leftarrow h_d$.
10. $e_d \leftarrow j$, $d \leftarrow d+1$, $j \leftarrow t+1$.
11. Return the sequence $(b_0, e_0), (b_1, e_1), \dots, (b_{d-1}, e_{d-1})$.

Notice that the algorithm scans the bits of k from right (least significant) to left, and as it progresses, it may modify (in Step 8) portions of the sequence $\{k_j\}$ that have not been processed yet. The correctness of the algorithm is verified inductively by asserting the condition

$$k = \sum_{i=0}^{d-1} b_i 2^{e_i} + \sum_{j'=j}^{\ell} k_{j'} 2^{j'} \tag{IV.7}$$

each time the loop condition in Step 2 is checked. Since the loop terminates with $j > \ell$, the second term of the sum in Equation (IV.7) vanishes, giving the desired decomposition of k. The proof is straightforward, the only key observation being that when the condition in Step 6 holds, Step 7 subtracts 2^{j+r} from the sum in Equation (IV.7) and Step 8 adds it back, since $t = j+r-1$ must hold in this case. Notice also that, by construction, all b_i produced are odd, and b_{d-1} must be positive when $k > 0$. Once the sequence $\{(b_i, e_i)\}_{i=0}^{d-1}$ is obtained, the point multiplication algorithm is a straightforward modification of the sliding window method. We assume $r > 1$, and $d \geq 1$ (i.e., $k > 0$).

ALGORITHM IV.7: **Point Multiplication: Signed m-ary Windows.**

```
INPUT:  A point, P, and {(b_i, e_i)}_{i=0}^{d-1} such that k = Σ_{i=0}^{d-1} b_i 2^{e_i}.
OUTPUT: Q = [k]P.
```

Precomputation.

```
1.  P_1 ← P, P_2 ← [2]P.
2.  For i = 1 to 2^{r-2}-1 do P_{2i+1} ← P_{2i-1} + P_2.
3.  Q ← P_{b_{d-1}}.
```

Main loop.

```
4.  For i = d-2 to 0 by -1 do:
5.      Q ← [2^{e_{i+1}-e_i}]Q.
6.      If b_i > 0 then Q ← Q + P_{b_i},
7.      Else Q ← Q - P_{-b_i}.
8.  Q ← [2^{e_0}]Q.
9.  Return Q.
```

Using an analysis similar to that of the unsigned sliding window scheme of Section IV.2.3, we can estimate the expected number of general point additions in the main loop of Algorithm IV.7 at $(n+1)/(r+1)-1$. The assumption of independence and uniform distribution of the bits k_j is more questionable here, since the modification of the sequence in Step 8 of Algorithm IV.6 does introduce a certain degree of dependency. However, the deviation is minimal, and the assumption, with respect to actual values used in practice, is not much worse than the original assumption of the input sequence k_j being uniformly distributed. On the other hand, the number of point operations in the precomputation phase is 2^{r-2}, i.e., about a half that of the unsigned method. Thus, the expected total number of point operations is estimated at

$$n + \frac{n+1}{r+1} + 2^{r-2} - 2. \qquad \text{(IV.8)}$$

Comparing this expression with the corresponding one for the NAF sliding window method in Equation (IV.4), we observe that the expression in Equation (IV.8) offers a trade off with more operations in the main loop (since $\nu(r) > 1$), but fewer operations in the precomputation phase. To bring the trade offs to a common comparison basis, we define r' so that $2^{r-2} = \frac{1}{3}2^{r'}$, i.e. $r' = r - (2 - \log_2 3)$. Then, Equation (IV.8) can be rewritten as

$$n + \frac{n+1}{r' + 3 - \log_2 3} + \frac{2^{r'}}{3} - 2. \tag{IV.9}$$

We conclude that the signed m-ary window method is asymptotically better than the windowed NAF method whenever $\nu(r) < 3 - \log_2 3 \approx 1.415$. This holds for all $r > 3$, by the expression for $\nu(r)$ in Equation (IV.3), which has $\nu(r) \to 4/3$ as $r \to \infty$. The margin of difference, however, is rather slim, and for practical values of n and r, once integer constraints and $O(1)$ terms are taken into account, the two schemes are very close in complexity.

IV.2.6. Example. The following example illustrates the different considerations and trade offs in the choice of a point multiplication algorithm.

Assume we need to compute $[k]P$, where

$$k = 741155629426723268099912038573.$$

The binary expansion of k, which is one hundred bits long and has weight 53, is given by

$$\underline{1001}\,0\,\underline{1011}\,0\,\underline{1011}\,00\,\underline{1101}\;\underline{1001}\,000\,\underline{101}\,00\,\underline{101}\,0\,\underline{1011}\,0\,\underline{111}\,00\,\underline{111}\,0\,\underline{1011}\,0\,\underline{1001}\,0\,\underline{11}\,00\,\underline{1101}\;\underline{11}\,00\backslash$$

$$\underline{1011}\,00\,\underline{101}\,000\,\underline{1111}\;\underline{1}\,000\,\underline{101}\,0\,\underline{1101}\,.$$

The underlined segments indicate the 'windows' processed by the unsigned sliding window method of Section IV.2.3, with $r = 4$. The number of such windows is 21. Therefore, the total number of curve operations for this method is

$$96(\text{doublings}) + 20(\text{additions}) + 8(\text{precomputation}) = 124.$$

The NAF of k has length 100 and weight 42, and it is given by

$$\underline{101}\,0\,\underline{\bar{1}0\bar{1}}\,00\,\underline{\bar{1}0\bar{1}}\,0\,\underline{\bar{1}01}\,00\,\underline{\bar{1}0\bar{1}}\,00\,\underline{1}\,000\,\underline{101}\,0\,\underline{10\bar{1}}\,0\,\underline{\bar{1}0\bar{1}}\,00\,\underline{\bar{1}}\,00\,\underline{\bar{1}01}\,000\,\underline{\bar{1}0\bar{1}}\,0\,\underline{\bar{1}01}\,0\,\underline{10\bar{1}}\,0\,\underline{\bar{1}01}\,00\,\underline{\bar{1}}\,00\,\underline{\bar{1}01}\backslash$$

$$0\,\underline{\bar{1}0\bar{1}}\,00\,\underline{101}\,00\,\underline{1}\,0000\,\underline{\bar{1}}\,00\,\underline{10\bar{1}}\,0\,\underline{\bar{1}0\bar{1}}\,0\,\underline{1}\,.$$

Here, the underlined segments indicate the windows processed by a sliding window method, applied to the NAF, with $r = 3$. The number of windows is 24. Hence, the total number of curve operations is

$$97(\text{doublings}) + 23(\text{additions}) + 3(\text{precomputation}) = 123.$$

(Curiously, for this value of k, the same number of curve operations is obtained with $r = 3, 4, 5$.)

The signed window m-ary decomposition of k, with $r = 5$, is given by the list of pairs

$$\begin{aligned}\{(b_i, e_i)\}_{i=0}^{18} \;=\; & \{(13,0),(5,5),(-1,11),(9,16),(-7,21),(-13,26),(7,33),\\ & (11,38),(13,44),(-3,49),(-3,54),(-9,59),(11,64),\\ & (5,70),(-7,76),(7,81),(11,86),(11,91),(9,96)\},\end{aligned}$$

satisfying $k = \sum_{i=0}^{18} b_i 2^{e_i}$, as can be readily verified using a suitable symbolic computation package. The total number of curve operations is

$$96(\text{doublings}) + 18(\text{additions}) + 8(\text{precomputation}) = 122.$$

Table IV.3 gives a more detailed analysis of the cost of computing $[k]P$ with various methods described in this chapter, in terms of field arithmetic operations. As usual, M indicates field multiplications and I field inversions. The table includes two lines for each method listed: in the first line, we assume that affine coordinates are used for all operations, while in the second line we assume that most operations are done in projective coordinates, with precomputations done in affine representation. In the latter case, the tally includes the cost of converting the final result back to affine coordinates. The columns under 'Total cost' give the cost of the computation, in M units, under two different assumptions of the relation between the costs of inversion and multiplication, namely $I = 3M$ and $I = 10M$. In each case, the lowest overall cost is indicated in boldface.

TABLE IV.3. Cost of point multiplication: an example.

Method	*Coordinates*	r	*Curve ops*	M	I	*Total cost* I=3M	I=10M
binary	affine	n/a	151	302	151	755	1812
	projective	n/a	151	1018	1	1021	1028
modified m-ary	affine	4	128	256	128	640	1536
	projective	4	128	739	9	766	829
sliding window	affine	4	124	248	124	620	1488
	projective	4	124	699	9	726	789
binary NAF	affine	n/a	140	280	140	700	1680
	projective	n/a	140	908	1	911	918
windowed NAF	affine	4	123	246	123	615	1476
	projective	3	123	724	4	736	**764**
signed m-ary	affine	5	**122**	244	122	**610**	1464
	projective	5	**122**	679	9	706	769

Table IV.3 confirms that affine coordinates are better when the ratio $I : M$ is relatively low, while projective coordinates are better when the ratio is high. The ratio depends strongly on the representations used, and on the computational environment. Examples of situations where the ratio might be high are a software implementation where the basic primitives of the multiplication routine have been tightly 'hand-coded' in machine language, or a hardware design containing a multiplier implementation but no inverter (recall that one can always realize inversion by means of multiplication). The table also shows signed window methods (NAF and m-ary) being superior to unsigned methods.

IV.2.7. Multiplying a fixed point. In some applications (e.g., part of the Diffie–Hellman key exchange protocol), we are required to compute multiples $[k]P$ of a *fixed* point P, known in advance of the computation. In such cases, a significant portion of the cost of point multiplication can be saved by precomputing and storing a table of multiples of P that is used for many values of k. For example, for the binary method, the multiples $[2^i]P$, $1 \leq i < \ell$, could be precomputed, eliminating all the doublings in the algorithm. Similar ideas can be used for m-ary and window methods. Various techniques for the general problem of fixed-basis exponentiation are described in [**48**] and [**99**].

IV.3. Frobenius Expansions

We say we are using a *subfield curve* when the group of rational points of interest (e.g., for implementing cryptographic protocols) is defined over a field $\mathbb{F}_{q^n}$, $n > 1$, but the coefficients of the curve are in $\mathbb{F}_q$. In this case, the multiplication procedure can be significantly accelerated by using a *Frobenius expansion.* In characteristic two this is based on ideas to be found in [**65**], [**96**], [**154**] and [**111**]. The idea also works in odd characteristic [**152**], where the trick in [**154**] for Euclidean endomorphism rings is also extended to non-Euclidean rings.

Notice that the concept of subfield curve is a relative one, in that it depends on the set of rational points we want to operate on, rather than on the curve itself. In some sense, all curves over finite fields are subfield curves.

Throughout, we let E denote an elliptic curve over the field $\mathbb{F}_q$, which we will implicitly assume to be small. For example, one can think of q as being less than 100. The extension $\mathbb{F}_{q^n}$ over which rational points are taken, on the other hand, is assumed to be large (a commonly used example is $q = 32$, $n = 31$ for rational points in $\mathbb{F}_{2^{155}}$).

We recall from Chapter III the qth-power Frobenius endomorphism,

$$\varphi : \begin{cases} E(\overline{\mathbb{F}}_q) & \longrightarrow & E(\overline{\mathbb{F}}_q) \\ (x, y) & \longmapsto & (x^q, y^q), \\ \mathcal{O} & \longmapsto & \mathcal{O}, \end{cases}$$

which satisfies the equation

$$\varphi^2 - [t]\varphi + [q] = [0].$$

Owing to the results in [**98**] and Chapter V, we shall assume that the curve is not supersingular, so the characteristic p does not divide the trace of Frobenius, $t = q + 1 - \#E(\mathbb{F}_q)$. By Hasse's Theorem we know that $|t| \leq 2\sqrt{q}$.

We can expand the multiplication map as a polynomial in φ, with 'small' coefficients and of bounded degree. As φ is easy to evaluate this greatly speeds up the multiplication operation. This is particularly noticeable if $\mathbb{F}_{q^n}$ is represented by a normal basis. In such a situation evaluation of φ in $\mathbb{F}_{q^n}$ is just a cyclic shift of the coefficients (over $\mathbb{F}_q$) of each point coordinate. For the rest of this chapter we justify this method and give explicit estimates on the size of such a Frobenius expansion.

To eliminate a few problem cases assume that $(q,t) \neq (5,\pm 4)$ or $(7,\pm 5)$ and $q \geq 4$. Such a restriction can be eliminated if some of the statements below are made more general. The method makes use of the fact that $\mathbb{Z}[\varphi]$ is a subring of $\mathrm{End}_{\mathbb{F}_q}(E)$ which is in turn isomorphic to a subring of $\mathbb{C}$. We will identify φ with its image under this isomorphism, treating it, when convenient, as a complex number satisfying the equation $\varphi^2 - t\varphi - q = 0$. We first show that an arbitrary element of $\mathbb{Z}[\varphi]$ can be trivially divided by φ to obtain a relatively small remainder.

LEMMA IV.2. *Let $S \in \mathbb{Z}[\varphi]$. Then there exist a unique integer $R \in \{-\lceil q/2 \rceil + 1, \ldots, \lfloor q/2 \rfloor\}$ and a unique element $Q \in \mathbb{Z}[\varphi]$ such that*

$$S = Q\varphi + R.$$

PROOF. Write $S = a + b\varphi$, with $a, b \in \mathbb{Z}$. Now write $a = Q'q + R$, with $Q' \in \mathbb{Z}$ and R in the desired range, and recall that $q = t\varphi - \varphi^2$. Then, $Q = b + Q't - \varphi$. □

The norm of an element $S = a + b\varphi \in \mathbb{Z}[\varphi]$ is defined to be

$$N_{\mathbb{Z}[\varphi]/\mathbb{Z}}(a + b\varphi) = a^2 + abt + b^2 q$$

We now show that an element in $\mathbb{Z}[\varphi]$ of small norm must have an expansion in φ which is short.

LEMMA IV.3. *Let $S \in \mathbb{Z}[\varphi]$ be such that*

$$N_{\mathbb{Z}[\varphi]/\mathbb{Z}}(S) \leq \begin{cases} (\sqrt{q}+1)^2, & q \text{ even}, \\ (\sqrt{q}+2)^2/4, & q \text{ odd}. \end{cases}$$

Then we can write

$$S = \sum_{i=0}^{3} a_i \varphi^i$$

with $a_i \in \{-\lceil q/2 \rceil + 1, \ldots, \lfloor q/2 \rfloor\}$.

PROOF. For a proof see [**111**] and [**152**]. □

We can now show that the Frobenius expansions exist and are of logarithmic length.

THEOREM IV.4. *Let $S \in \mathbb{Z}[\varphi]$. Then, we can write*

$$S = \sum_{i=0}^{k} r_i \varphi^i$$

where $r_i \in \{-\lceil q/2 \rceil + 1, \ldots, \lfloor q/2 \rfloor\}$ and $k \leq \lceil 2\log_q 2\sqrt{N_{\mathbb{Z}[\varphi]/\mathbb{Z}}(S)} \rceil + 3$.

PROOF. From Lemma IV.2 we can obtain an expansion of the form

$$\begin{aligned} S &= S_0 = S_1\varphi + r_0 = (S_2\varphi + r_1)\varphi + r_0 \\ &= \sum_{i=0}^{j} r_i\varphi^i + S_{j+1}\varphi^{j+1} \end{aligned}$$

Using the triangle inequality we see, defining $\|\cdot\| = \sqrt{N_{\mathbb{Z}[\varphi]/\mathbb{Z}}(\cdot)}$, the following:

(i) q even,

$$\begin{aligned} \|S_{j+1}\| &\leq \frac{\|S_j\| + \|r_i\|}{\|\varphi\|} \leq \frac{\|S_j\| + q/2}{\sqrt{q}} = \frac{\|S_0\|}{q^{(j+1)/2}} + \frac{1}{2}\sum_{i=-1}^{j-1} q^{-i/2} \\ &\leq \frac{\|S_0\|}{q^{(j+1)/2}} + \sqrt{q}. \end{aligned}$$

(ii) q odd,

$$\begin{aligned} \|S_{j+1}\| &\leq \frac{\|S_j\| + \|r_i\|}{\|\varphi\|} \leq \frac{\|S_j\| + (q-1)/2}{\sqrt{q}} = \frac{\|S_0\|}{q^{(j+1)/2}} + \frac{(q-1)}{2}\sum_{i=1}^{j+1} q^{-i/2} \\ &\leq \frac{\|S_0\|}{q^{(j+1)/2}} + \frac{\sqrt{q}+1}{2}. \end{aligned}$$

In both cases, if $j \geq \lceil 2\log_q 2\|S_0\| \rceil - 1$ then

$$\frac{\|S_0\|}{q^{(j+1)/2}} \leq 1/2.$$

Hence

$$N_{\mathbb{Z}[\varphi]/\mathbb{Z}}(S_{j+1}) \leq \begin{cases} (\sqrt{q}+1)^2, & q \text{ even}, \\ (\sqrt{q}+2)^2/4, & q \text{ odd}, \end{cases}$$

and so by Lemma IV.3 we know that S_{j+1} has a Frobenius expansion of length at most 4. □

To use this to implement multiplication by m on the elliptic curve, first consider m as an element of $\mathbb{Z}[\varphi]$ and compute its Frobenius expansion,

$$m = \sum_{i=0}^{k} r_i\varphi^i,$$

where $k \leq \lceil 2\log_q 2m \rceil + 3$. The points $[m]P$ for $P \in E(\mathbb{F}_{q^n})$ can then be computed using Horner's method:

$$\begin{aligned} [m]P &= \sum_{i=0}^{k} [r_i]\varphi^i(P) \\ &= \varphi(\ldots\varphi([r_k]\varphi(P) + [r_{k-1}]P) + \cdots + [r_1]P) + [r_0]P. \end{aligned}$$

Note at each stage of the expansion an element of the form $[r]P$, where $|r| \leq \lfloor q/2 \rfloor$, is added. To speed up this step a table of such multiples could be precomputed. This would be particularly useful if many multiplications of the same point were required.

The length of the Frobenius expansion can be reduced by nearly a half using a trick of Solinas [**154**]. To explain this a small generalization of the notion of Euclidean domains is required:

DEFINITION IV.5. *Let λ be a positive real number, let A denote a commutative ring, and suppose that there exists a multiplicative function*

$$\Psi : A \setminus \{0\} \to \mathbb{N}.$$

The ring will be called λ-Euclidean if for all $a, b \in A$, with $b \neq 0$, we can find $q, r \in A$ with

$$a = bq + r$$

such that either $r = 0$ or $\Psi(r) < \lambda\Psi(b)$.

Such an idea is not new as one can see by looking at the survey article [**75**]. Suppose A has field of fractions K. Then, we can extend Ψ to $K \setminus \{0\}$ in the obvious way. We then have

LEMMA IV.6. *The ring A will be λ-Euclidean if for all $x \in K$ we can find a $y \in A$ such that*

$$\Psi(x - y) < \lambda.$$

The main result on λ-Euclidean rings that will be used is the following.

THEOREM IV.7. *Suppose $\varphi^2 - t\varphi + q = 0$. Then, $\mathbb{Z}[\varphi]$ is λ-Euclidean for some λ such that $0 < \lambda \leq (9 + 4q)/4$.*

The proof of the theorem is straightforward (see [**152**]). This result is used to reduce the length of the Frobenius expansion by around 50%. Consider the integer, m, we wish to multiply P by as being an element of $\mathbb{Z}[\varphi]$. As $m \approx q^n$, the norm of m will be equal to m^2 which is of order approximately q^{2n}. However, for points $P \in E(\mathbb{F}_{q^n})$ we have the identity

$$\varphi^n P = P.$$

So m can be 'divided' by $\varphi^n - 1$ to obtain a remainder r with

$$N_{\mathbb{Q}[\varphi]/\mathbb{Q}}(r) < \lambda N_{\mathbb{Q}[\varphi]/\mathbb{Q}}(\varphi^n - 1) \leq \frac{9 + 4q}{4} N_{\mathbb{Q}[\varphi]/\mathbb{Q}}(\varphi^n - 1) \approx q^{n+1}.$$

Hence, multiplication by m can be replaced by multiplication by r. As r has norm roughly q^{n+1}, its Frobenius expansion will be nearly half the length of the Frobenius expansion of m. This should provide a 50% improvement in the performance of the algorithm.

IV.4. Point Compression

In cryptographic protocols based on elliptic curves, it is often necessary to store (e.g., in a public key directory) or transmit (e.g., in a Diffie–Hellman key exchange) elliptic curve points. When resources such as storage or bandwidth are at a premium, it is desirable to represent those points using the minimum possible number of bits. This is referred to as *point compression*.

In a full representation, an affine point (x,y) requires $2n$ bits, where n is the extension degree if working in $\mathbb{F}_{2^n}$, or $n = \lceil \log_2 p \rceil$ if working in $\mathbb{F}_p$. The number of bits is trivially reduced to $n+1$ by observing that given the x-coordinate of a curve point, the elliptic curve equation becomes a quadratic in y. Therefore, one bit, used to distinguish between the (at most) two solutions of the quadratic equation, is sufficient to specify y. Decompressing a point so specified involves solving the quadratic equation, which can be done using the techniques described in Chapter II.

In the case of $\mathbb{F}_{2^n}$, Seroussi [**145**] observes that an additional bit can be saved in the x-coordinate, when the curve point has odd order. This is the case in the applications mentioned above, where all the points involved belong to a large subgroup of prime order, over which the ECDLP is defined. The savings derive from the following lemma.

LEMMA IV.8. *Let $P = (x, y)$ be a rational point of odd order on the curve*

$$E : Y^2 + XY = X^3 + a_2X^2 + a_6,$$

over $\mathbb{F}_{2^n}$. Then,

$$\mathrm{Tr}_{q|2}(x) = \mathrm{Tr}_{q|2}(a_2). \tag{IV.10}$$

PROOF. If $P \in E(\mathbb{F}_{2^n})$ has odd order, then $P = [2]Q$ for some point $Q = (x_1, y_1) \in E(\mathbb{F}_{2^n})$. From the point doubling formula in Section IV.1.2, we have $x = \lambda^2 + \lambda + a_2$, where $\lambda = x_1 + y_1/x_1$. Thus, $\mathrm{Tr}_{q|2}(x) = \mathrm{Tr}_{q|2}(a_2)$. □

Since the parameters of the curve are assumed known by all communicating parties, Equation (IV.10) poses a fixed linear constraint on x. It follows that $n-1$ bits are sufficient to specify x, and n bits are sufficient to fully specify a point in the subgroup of interest.

One might ask if it is possible to represent a point with fewer than n bits. For some values of n, and a system designed to support all possible values of a_2 and a_6, n bits are indeed necessary as we will now show. By Hasse's Theorem, the order of the group falls in the range $2^n+1-2\sqrt{2^n} \le \#E(\mathbb{F}_{2^n}) \le 2^n+1+2\sqrt{2^n}$. The order also satisfies $\#E(\mathbb{F}_{2^n}) \equiv 2b \pmod 4$, where $b=1$ if $\mathrm{Tr}_{q|2}(a_2) = 1$ and 0 otherwise (see Lemma III.4). Thus, when $\mathrm{Tr}_{q|2}(a_2) = 1$, the size of the largest prime subgroup of $E(\mathbb{F}_{2^n})$ can reach $\#E(\mathbb{F}_{2^n})/2$ and exceed 2^{n-1}, requiring n bits to specify a subgroup element. This situation arises, for instance, with the curve of Example 12 in Appendix A. The curve is defined over $\mathbb{F}_{2^{131}}$, and its group of rational points is of size $\#E(\mathbb{F}_{2^{131}}) = 2p$, where p is the prime

$$p = 2^{130} + 1117721673928288 7043.$$

Similarly, the curves of Examples 14, 16, 18, 21, and 22 in the appendix, defined over $\mathbb{F}_{2^n}$ with $n = 163, 191, 239, 367$, and 401, respectively, all have groups of rational points with orders of the form $2p$, where (for each respective value of n) p is a prime satisfying $p > 2^{n-1}$.

The question of whether an infinite sequence of values of n exists for which such curves can be found is open, and related to the (hard) question of whether there is an infinite sequence of primes in the set

$$\bigcup_j \left\{ i \mid 2^j+1-2\sqrt{2^j} \leq 2i \leq 2^j+1+2\sqrt{2^j} \right\};$$

see [64] and Section VI.5.

When $\mathrm{Tr}_{q|2}(a_2) = 0$, $n-1$ bits could suffice to represent a subgroup point, as $p \leq \#E(\mathbb{F}_{2^n})/4$ in this case. However, no efficient method to obtain such a representation is known.

CHAPTER V

The Elliptic Curve Discrete Logarithm Problem

Let E be an elliptic curve over some finite field, $\mathbb{F}_q$. In what follows let n denote the order of the group $E(\mathbb{F}_q)$ and let P denote an element of $E(\mathbb{F}_q)$. The elliptic curve discrete logarithm problem (ECDLP) on E is, given $Q \in \langle P \rangle$, find the integer, m, such that

$$Q = [m]P.$$

There are a number of ways of approaching the solution to this problem. The first two listed below depend on the fact that the group of interest is the group of points of an elliptic curve, or a prime order subgroup. These approaches will correspond to the MOV and anomalous attacks respectively, referred to in Chapter III and discussed in Sections V.2 and V.3. The final two approaches do not make any explicit assumption about the underlying group. Methods like the final two are often referred to as being applicable to 'black box groups', and in some sense are the best possible for that class of groups [**146**]. Complexity will be measured in terms of the number of basic group operations, comparisons etc. which need to be performed rather than the bit complexity.

The methods to be covered in this chapter are:

1. Using a Weil pairing on $E[n]$, there is a polynomial time reduction, in terms of the number of operations in $\mathbb{F}_{q^l}$, of the ECDLP on $E(\mathbb{F}_q)$ to the DLP in $\mathbb{F}_{q^l}$, for some integer l (see [**98**] and [**44**]). The integer l that is required is the smallest such that $q^l \equiv 1 \pmod{n}$, when $\gcd(n, q) = 1$. This is the MOV attack.
2. Suppose now that q is a prime. For anomalous elliptic curves (trace of Frobenius $t = 1$, and $n = q$), by using the q-adic elliptic logarithm one can give a linear time method to solve the ECDLP (see [**153**] and [**136**]). This method is related to another linear time method of Semaev [**143**] to solve the ECDLP in the q-primary part of the subgroup generated by P. Anomalous curves had been proposed for use by Miyaji [**104**] as they are particularly able to resist the MOV attack.
3. The BSGS method of Shanks can be used to solve the DLP in any finite abelian group. This method is a standard time/memory trade off and has space and time complexity given by $O(\sqrt{n})$.
4. Using random walks one can reduce the space to a constant amount and still maintain a time complexity of $O(\sqrt{n})$. This is done using

one of two strategies, both due to Pollard, called the *rho* and *lambda* methods. The second of these is often referred to as the method of *tame and wild kangaroos.* Another advantage of the random walks method is that it can be efficiently parallelized [**116**].

It should be noted that none of the attacks listed above will be effective for a carefully chosen elliptic curve. To avoid the MOV attack it is important to choose an elliptic curve group order that does not divide $q^l - 1$ for small values of l (this is quantified in Section V.7). It will be observed in Section V.2 that it is not possible to do this for supersingular curves, which is the reason they are not considered appropriate for cryptographic applications. Anomalous curves are a very small class of curves, and are avoided because of the second attack mentioned. The last two attacks listed are quite general but have complexities on the order of $\sqrt{n}$. They become infeasible when the curve order is sufficiently large.

Before discussing these methods we reduce the problem to one of prime order subgroups. Throughout the rest of the chapter, we use elliptic curve additive notation for the group G, although some of the methods described apply to general finite abelian groups.

V.1. The Simplification of Pohlig and Hellman

Pohlig and Hellman [**124**] noticed that to solve the DLP in a finite abelian group G one need only solve the DLP in subgroups of prime power order in G. The original DLP is then solved by appealing to the *Chinese Remainder Theorem* (CRT). In addition, the problem can be reduced to the case of prime order subgroups, as will now be shown. An obvious consequence of this fact is that to maintain security of a system based on the DLP, the order of G should contain a large prime divisor. Here, by 'large' we mean large enough to preclude solving a DLP in the prime order subgroup.

Let G have order divisible by a prime p and suppose we wish to solve the following DLP:

$$Q = [m]P.$$

If G has order n, then the problem can be restricted to a subgroup of order p by solving

$$Q' = [n']Q = [m_0]([n']P) = [m_0]P'$$

where $n' = n/p^{c-1}$, p^c is the largest power of p dividing n. Thus P' is a point of order p. Solving this problem will determine the value, m_0, of m modulo p.

The values of m modulo $p^2, p^3, \ldots, p^c$ are then computed in the following way. Suppose $m \equiv m_i \pmod{p^i}$ is known and $m = m_i + \lambda p^i$ for some integer $\lambda \in \mathbb{Z}$. Then

$$R = (Q - [m_i]P) = [\lambda]([p^i]P) = [\lambda]S,$$

where R and S are known and S has order $s = n/p^i$. The value of $\lambda \pmod p$ can be determined just as $m \pmod p$ was found above. Let $s' = s/p^{c-i-1}$. Then, $\lambda \pmod p$ is obtained by solving the DLP

$$R' = [s']R = [\lambda_0]([s']S) = [\lambda_0]S',$$

where S' is a point of order p.

Continuing in this manner, by solving DLPs in subgroups of order p, we eventually determine m modulo p^c. After computing m modulo p^c for all prime divisors p of n, the true solution, m, to the original DLP can be obtained using the CRT.

V.1.1. Example. As an example of this method consider the elliptic curve

$$E : Y^2 = X^3 + 71X + 602$$

over the finite field $\mathbb{F}_{1009}$. The group order of $E(\mathbb{F}_{1009})$ is 1060 which is $2^2 \cdot 5 \cdot 53$. Suppose the two points

$$P = (1, 237) \text{ and } Q = (190, 271)$$

are given and the solution to the ECDLP

$$Q = [m]P.$$

is required. First notice that P has order $530 = 2 \cdot 5 \cdot 53$ in the group $E(\mathbb{F}_{1009})$. Hence by the above reduction of Pohlig and Hellman, the computation of m can be reduced to the computation of m modulo 2, 5 and 53.

The solution modulo 2. By the above method we need to multiply P and Q by $530/2 = 265$ to obtain points of order 2. The ECDLP in the subgroup of order 2 can then be solved and hence m modulo 2 deduced. It is found that

$$\begin{aligned} P_2 &= [265]P = (50, 0), \\ Q_2 &= [265]Q = (50, 0). \end{aligned}$$

The ECDLP becomes

$$Q_2 = [m \pmod 2)]P_2,$$

and it is deduced that $m \equiv 1 \pmod 2$.

The solution modulo 5. The points are multiplied by $530/5 = 106$, to obtain

$$\begin{aligned} P_5 &= [106]P = (639, 160), \\ Q_5 &= [106]Q = (639, 849). \end{aligned}$$

Hence $Q_5 = -P_5$ and $m \equiv 4 \pmod 5$.

The solution modulo 53. The points are multiplied by $530/53 = 10$ to obtain

$$\begin{aligned} P_{53} &= [10]P = (32, 737) = P', \\ Q_{53} &= [10]Q = (592, 97) = Q'. \end{aligned}$$

Clearly in this example the value of $m \pmod{53}$ could be determined using a brute force search. However, we instead use the ECDLP

$$Q' = [m_0]P'$$

to illustrate the baby step/giant step method and the random walks method in a later section.

V.2. The MOV Attack

We shall now explain the method of Frey and Rück [**44**] (as described by Voloch [**161**]) which generalizes the result of Menezes, Okamoto and Vanstone [**98**] (usually referred to as MOV). The explanation will be dependent on the theory of descents, to make it immediate to those who come from a background in the number-theoretic study of elliptic curves.

The description given is rather vague but it makes the main points clear. After skimming over the mathematics we give a more down to earth description of what are the actual steps required. Much of the notation and argument related to descents can be found in [**147**].

V.2.1. Descent via isogeny. Let E denote an elliptic curve defined over a field $\mathbb{F}_q$ and let n denote a prime number, which is coprime to q. Let $\mathbb{F}_{q^l}$ denote the field which is obtained by adjoining the nth roots of unity to $\mathbb{F}_q$. Assume an isogeny

$$\phi : E' \longrightarrow E$$

is given, whose kernel has exponent n, and the points of the kernel of ϕ are defined over $\mathbb{F}_{q^l}$.

The standard exact sequence of Galois modules is then

$$0 \longrightarrow E'[\phi] \longrightarrow E'(\overline{\mathbb{F}}_q) \longrightarrow E(\overline{\mathbb{F}}_q) \longrightarrow 0,$$

and taking Galois cohomology we find the long exact sequence, setting $G = \mathrm{Gal}(\overline{\mathbb{F}}_{q^l}/\mathbb{F}_{q^l})$,

$$\begin{array}{cccccccc} 0 & \longrightarrow & E'(\mathbb{F}_{q^l})[\phi] & \longrightarrow & E'(\mathbb{F}_{q^l}) & \longrightarrow & E(\mathbb{F}_{q^l}) & \\ & \longrightarrow & H^1(G, E'[\phi]) & \longrightarrow & H^1(G, E'(\overline{\mathbb{F}}_{q^l})) & \longrightarrow & \cdots & . \end{array}$$

The following short exact sequence can be deduced from this, which is the main sequence in the theory of descents for elliptic curves:

$$0 \longrightarrow E(\mathbb{F}_{q^l})/\phi(E'(\mathbb{F}_{q^l})) \longrightarrow H^1(G, E'[\phi]) \longrightarrow H^1(G, E')[\phi] \longrightarrow 0.$$

Now since $E'[\phi] \subset E'(\mathbb{F}_{q^l})$ the action of G on $E'[\phi]$ is trivial and so we have $H^1(G, E'[\phi]) = Hom(G, E'[\phi])$, the group of homomorphisms from G to $E'[\phi]$. The first non-trivial arrow in the last exact sequence is given by

$$\delta_E : \begin{cases} E(\mathbb{F}_{q^l})/\phi E'(\mathbb{F}_{q^l}) & \longrightarrow & Hom(G, E'[\phi]) \\ \mathcal{P} & \longmapsto & \sigma \longmapsto \mathcal{Q}^\sigma - \mathcal{Q} \end{cases}$$

where $\mathcal{P} \in E(\mathbb{F}_{q^l})$ and $\mathcal{Q} \in E'(\overline{\mathbb{F}}_{q^l})$ is chosen so that $[\phi]\mathcal{Q} = \mathcal{P}$.

Since $\mathbb{F}_{q^l}$ contains the group of nth roots of unity, denoted by $\mu_n(\mathbb{F}_{q^l})$, by Hilbert's Theorem 90, we have an isomorphism

$$\delta_K : \begin{cases} \mathbb{F}_{q^l}^*/(\mathbb{F}_{q^l}^*)^n & \longrightarrow & Hom(G, \mu_n) \\ b & \longmapsto & \sigma \longmapsto \beta^\sigma/\beta \end{cases}$$

where $b \in \mathbb{F}_{q^l}^*$, $\beta \in \overline{\mathbb{F}}_{q^l}$ is chosen so that $\beta^n = b$, and $\mathbb{F}_{q^l}^*/(\mathbb{F}_{q^l}^*)^n$ denotes the quotient group of $\mathbb{F}_{q^l}^*$ modulo the nth powers of elements in $\mathbb{F}_{q^l}^*$. Given the above definitions it is then a standard fact that there exists a bilinear pairing, κ, which is non-degenerate on the left,

$$\kappa : \begin{cases} E(\mathbb{F}_{q^l})/\phi E'(\mathbb{F}_{q^l}) \times E[\hat{\phi}] & \longrightarrow & \mathbb{F}_{q^l}^*/(\mathbb{F}_{q^l}^*)^n \\ (\mathcal{P}, \mathcal{T}) & \mapsto & \delta_K^{-1}(e_\phi(\delta_E(\mathcal{P}), \mathcal{T})), \end{cases}$$

where $e_\phi(R, S)$ is the pairing from Lemma III.13. It follows that

$$\kappa(\mathcal{P}, \mathcal{T}) \equiv f_{\mathcal{T}}(\mathcal{P}) \pmod{(\mathbb{F}_{q^l}^*)^n}$$

for some computable function on the curve, $f_{\mathcal{T}}$, defined over $\mathbb{F}_{q^l}$. The function $f_{\mathcal{T}}(\mathcal{P})$ is computed in much the same way as the Weil pairing is computed (see [**147**, Theorem X.1.1], [**138**] and below).

For later use note that the groups $\mathbb{F}_{q^l}^*/(\mathbb{F}_{q^l}^*)^n$ and $\mu_n(\mathbb{F}_{q^l})$ are isomorphic via the isomorphism

$$\Upsilon : \begin{cases} \mathbb{F}_{q^l}^*/(\mathbb{F}_{q^l}^*)^n & \longrightarrow & \mu_n(\mathbb{F}_{q^l}) \\ \alpha \cdot (\mathbb{F}_{q^l}^*)^n & \longmapsto & \alpha^{(q^l-1)/n}. \end{cases}$$

V.2.2. The reduction. Consider solving the ECDLP

$$Q = [m]P,$$

where P is a point of order n in $E(\mathbb{F}_q)$. By the simplification of Pohlig and Hellman we can assume that n is prime. Choose l as before to be the minimal integer such that $\mathbb{F}_{q^l}$ contains the nth roots of unity, in particular $q^l \equiv 1 \pmod{n}$.

For supersingular curves over prime fields we have $n = p + 1$, therefore $p^2 = n^2 - 2n + 1 \equiv 1 \pmod{n}$, and so l can be chosen as 2. It can be shown [**98**] for supersingular curves, where $\mathrm{char}(\mathbb{F}_q)$ divides the trace of Frobenius t, that t^2 can only take on the values 0, q, $2q$, $3q$ and $4q$. Thus, the elliptic curve group orders are restricted to $q + 1 \pm \sqrt{jq}$, $j = 0, 1, 2, 3, 4$, and it is straightforward to verify that these orders divide at least one of the values

$q^l - 1$ for $l \leq 6$. For curves of trace two over prime fields the situation is even worse, since $n = p - 1$ and so $l = 1$.

There are two cases to consider, the first is when $l > 1$ and the second is when $l = 1$. In the following discussion we let $\mathbb{F}_q(E[n])$ denote the field obtained by adjoining the x- and y-coordinates of all the points of order n of E to the field $\mathbb{F}_q$.

Case $l > 1$. We have $\mathbb{F}_{q^l} = \mathbb{F}_q(E[n])$, by Lemma III.9 and the definition of l. If in the above discussion $E = E'$ and ϕ is taken to be the multiplication-by-n map, then the pairing

$$\kappa : \left\{ \begin{array}{ccc} E(\mathbb{F}_{q^l})/nE(\mathbb{F}_{q^l}) \times E[n] & \longrightarrow & \mathbb{F}_{q^l}^*/\mathbb{F}_{q^l}^{*\,n} \\ (\mathcal{P}, \mathcal{T}) & \mapsto & f_{\mathcal{T}}(\mathcal{P}) \end{array} \right.$$

is obtained. Choosing $\mathcal{T} \in E[n] \setminus E(\mathbb{F}_q)$ we obtain that $\Upsilon(f_{\mathcal{T}}(P))$ has exact order n and the map

$$\Psi : \left\{ \begin{array}{ccc} \langle P \rangle & \longrightarrow & \mu_n(\mathbb{F}_{q^l}) \\ \mathcal{P} & \mapsto & \Upsilon(f_{\mathcal{T}}(\mathcal{P})) \end{array} \right.$$

is an injection. Hence to solve an ECDLP in $\langle P \rangle$ we need only map the problem over to $\mu_n(\mathbb{F}_{q^l})$ and solve it there using one of the known sub-exponential methods. Clearly this will only be of advantage if l is relatively small.

Case $l = 1$. Here it may not be true that $\mathbb{F}_{q^l} = \mathbb{F}_q(E[n])$ and a little more care is needed. By Theorem III.11, there is an elliptic curve E' defined over $\mathbb{F}_q$ and an isogeny

$$\hat{\phi} : E \longrightarrow E'$$

with kernel $\langle P \rangle$. So in our descent discussion above take ϕ to be the dual to $\hat{\phi}$. By Lemma III.13, since $\mathbb{F}_q$ contains the nth roots of unity, the points of the kernel of ϕ are also defined over $\mathbb{F}_q$, and hence all the conditions imposed at the beginning of the above discussion hold. The pairing

$$\kappa : \left\{ \begin{array}{ccc} E(\mathbb{F}_q)/\phi E'(\mathbb{F}_q) \times \langle P \rangle & \longrightarrow & \mathbb{F}_q^*/\mathbb{F}_q^{*n} \\ (\mathcal{P}, \mathcal{T}) & \mapsto & f_{\mathcal{T}}(\mathcal{P}) \end{array} \right.$$

is thus obtained. Now as $E(\mathbb{F}_q)/\phi E'(\mathbb{F}_q)$ contains a subgroup isomorphic to $\langle P \rangle$, if we choose $\mathcal{T} = P$ then $\Upsilon(f_{\mathcal{T}}(P))$ is an element of exact order n. This last fact follows from the non-degeneracy of the ϕ-Weil pairing. The injection

$$\Psi : \left\{ \begin{array}{ccc} \langle P \rangle & \longrightarrow & \mu_n(\mathbb{F}_q) \\ \mathcal{P} & \mapsto & \Upsilon(f_{\mathcal{T}}(\mathcal{P})) \end{array} \right.$$

is then obtained, and so we can solve the ECDLP in $\langle P \rangle$ by mapping it over to $\mu_n(\mathbb{F}_q)$.

V.2.3. Description of $f_{\mathcal{T}}(\mathcal{P})$. To construct the map $f_{\mathcal{T}}$ a point $\mathcal{T}$ is chosen as above. In practice this is an element of exact order n which lies in $E(\mathbb{F}_{q^l})$. If $l \neq 1$ then we insist that $\mathcal{T}$ does not lie in $E(\mathbb{F}_q)$. Constructing such a $\mathcal{T}$ can be done by determining a random point, $\mathcal{S} \in E(\mathbb{F}_{q^l})$, and then multiplying it by $\#E(\mathbb{F}_{q^l})/n$ to obtain $\mathcal{T}$. With high probability we obtain $\mathcal{T} \neq \mathcal{O}$. However if $\mathcal{T} = \mathcal{O}$ then this value is rejected and another random point $\mathcal{S}$ is chosen. Hence a value of $\mathcal{T}$ is easily obtained which is not equal to $\mathcal{O}$. Since n is prime, such a point will have exact order n.

A second piece of information to $\mathcal{T}$ is added which is initially set to one, hence $\mathcal{T} = ((x, y), 1)$. The point $[n]\mathcal{T}$ is computed using an addition chain (or binary) method using the following modified addition procedure:

ALGORITHM V.1: **Modified Addition Algorithm**

INPUT: Two points (P_1, f_1), (P_2, f_2) with $P_i \in E[n]$.
OUTPUT: The sum (P_3, f_3).

1. Set $P_3 \leftarrow P_1 + P_2$ using the usual addition formulae.
2. Let $l(X,Y) = 0$ denote the equation of the line going through the points P_1 and P_2.
3. Let $v(X,Y) = 0$ denote the equation of the line through P_3 and $\mathcal{O}$, using the constant 1 in lieu of $v(X,Y)$ if $P_3 = \mathcal{O}$.
4. $f_3 \leftarrow f_1 f_2 l(X,Y)/v(X,Y)$.
5. Return (P_3, f_3).

Upon computing $[n]\mathcal{T} = (\mathcal{O}, f_{\mathcal{T}})$ it follows that $f_{\mathcal{T}}$ is the required function. Clearly there is no need to actually compute the function $f_{\mathcal{T}}$ as a rational function in $\mathbb{F}_{q^l}(X, Y)$, since the above method to evaluate $f_{\mathcal{T}}$ at any point $P = (x, y)$ can be used by substituting the values of x and y for the indeterminates X and Y in the algorithm.

However, this function, $f_{\mathcal{T}}$, is only defined on 'good' divisors. A divisor is good if no element in its support is equal to any of the multiples of $\mathcal{T}$ which occur in the addition chain for computing $f_{\mathcal{T}}$. So, for example, a divisor is good if its support is distinct from $\langle \mathcal{T} \rangle$.

There is an isomorphism of an elliptic curve with its divisor class group which to a point $\mathcal{P}$ associates the divisor class containing the divisor

$$(\mathcal{P}) - (\mathcal{O}). \tag{V.1}$$

In the case of curves of the form

$$Y^2 = X^3 + AX + B,$$

a good divisor can be found which is equivalent in the divisor class group to the divisor in Formula (V.1) by finding an $\mathbb{F}_q$-rational point S (which does not need to lie on the curve), and looking at the line which passes through S and $-\mathcal{P}$. This line should not pass through any point in $\langle \mathcal{T} \rangle$, other than possibly $-\mathcal{P}$. Then let P_1 and P_2 denote the two other points of intersection of this line with the curve. Let a denote an element in the field, $\mathbb{F}_q$, which is not the x-coordinate of a point in $\langle \mathcal{T} \rangle$. Define $Q_1 = (a, b)$ and $Q_2 = (a, -b)$ (it does not matter that b is not in $\mathbb{F}_q$ as the set $\{Q_1, Q_2\}$ will be $\mathbb{F}_q$-rational). Then, we have the following representation in the divisor class group:

$$(\mathcal{P}) - (\mathcal{O}) \equiv (P_1) + (P_2) - (Q_1) - (Q_2),$$

where the divisor on the right is $\mathbb{F}_q$-rational and has support distinct from $\langle \mathcal{T} \rangle$.

Alternatively, we may be able to write

$$(\mathcal{P}) - (\mathcal{O}) \equiv ([a+1]\mathcal{P}) - ([a]\mathcal{P}),$$

if $[a+1]\mathcal{P}$ and $[a]\mathcal{P}$ do not arise in the definition of $f_{\mathcal{T}}$. This will then be adequate as a good divisor to apply the function $f_{\mathcal{T}}$. This alternative can be used for fields of characteristic two.

Before passing on it should be noted that the papers of Menezes et al. [**98**] and Frey and Rück [**44**] contain further improvements and methods related to the above attack. In addition, there is one class of elliptic curves for which the above method cannot be applied no matter how large a value of l we take. These are the 'anomalous' curves over $\mathbb{F}_p$, which are ones such that $\#E(\mathbb{F}_p) = p$. To see that no integer $l \geq 1$ exists for these curves we notice that

$$p^l \equiv 0 \not\equiv 1 \pmod{p}.$$

This was noticed by Miyaji [**104**], who proposed such curves for use in cryptography for exactly this reason. However, as will be shown in the next section, such curves are very weak but for an entirely different reason.

The MOV attack will first be illustrated with three examples of discrete logarithm problems on elliptic curves over prime fields of odd characteristic. Two examples are for curves of trace two, and one is for a curve of trace zero (and hence supersingular).

V.2.4. Example 1. Consider first the following example. Let E denote the elliptic curve, defined over $\mathbb{F}_{173}$,

$$E : Y^2 = X^3 + 146X + 33,$$

which has trace 2 and hence order 172. An element of order 43 is given by $P = (168, 133)$. The solution of the ECDLP given by $Q = [m]P$, where $Q = (147, 74)$ is required.

Take $\mathcal{T} = (168, 133)$, which has order 43, and write $(P) - (\mathcal{O})$ and $(Q) - (\mathcal{O})$ as the 'good' divisors:

$$\begin{aligned}(P) - (\mathcal{O}) &= ([10]P) - ([9]P),\\ (Q) - (\mathcal{O}) &= ([10]Q) - ([9]Q).\end{aligned}$$

Note that none of $[10]P, [9]P, [10]Q$ and $[9]Q$ appears in the binary algorithm required to multiply $\mathcal{T}$ by 43. Then evaluate $\Psi = \Upsilon \circ f_{\mathcal{T}}$ at these four points and compute

$$\Psi(P) = \frac{\Psi([10]P)}{\Psi([9]P)} = 81,$$
$$\Psi(Q) = \frac{\Psi([10]Q)}{\Psi([9]Q)} = 139.$$

It is then seen that

$$81^{19} \equiv 139 \pmod{173}$$

and it is easily checked that 19 is a solution to our DLP on the elliptic curve.

V.2.5. Example 2. We consider the supersingular elliptic curve over $\mathbb{F}_{151}$ defined by

$$E : Y^2 = X^3 + 2X.$$

This curve has order 152. An element of order 19 in $E(\mathbb{F}_{151})$ is given by $P = (97, 26)$ and the solution to the DLP given by

$$Q = (43, 4) = [m]P$$

is sought. Notice that $151^2 \equiv 1 \pmod{19}$ and so computations are done in $E(\mathbb{F}_{151^2})$. Set

$$K = \mathbb{F}_{151^2} = \mathbb{F}_{151}[\theta]/(\theta^2 + 31\theta + 70).$$

An element of order 19 in $E(K) \setminus E(\mathbb{F}_{151})$ is given by

$$\mathcal{T} = (115\theta + 142, 141\theta + 86).$$

Since $\langle \mathcal{T} \rangle$ has a trivial intersection with $\langle P \rangle$ take $([2]P) - (P)$ and $([2]Q) - (Q)$ to be the good divisors equivalent to $(P) - (\mathcal{O})$ and $(Q) - (\mathcal{O})$. Then compute $\Psi = \Upsilon \circ f_{\mathcal{T}}$ as before, to obtain

$$\Psi(P) = \left(\frac{f_{\mathcal{T}}([2]P)}{f_{\mathcal{T}}(P)} \right)^{1200} = 44\theta + 102 = \alpha,$$
$$\Psi(Q) = \left(\frac{f_{\mathcal{T}}([2]Q)}{f_{\mathcal{T}}(Q)} \right)^{1200} = 9\theta + 100 = \beta.$$

The DLP

$$\beta = \alpha^m$$

can then be solved in K to determine that $m = 10$. Hence the solution to the ECDLP on our elliptic curve is also 10.

V.2.6. Example 3. This section is concluded with an example which may help illustrate some of the other points above. Take

$$E : Y^2 = X^3 + 16X + 27,$$

over the field $\mathbb{F}_{29}$. This has trace 2 and hence order 28. An element of order 7 is given by $\mathcal{T} = (21, 24)$. Compute the function $f_{\mathcal{T}}$ using the algorithm above to find

$$f_{\mathcal{T}}(X, Y) = \frac{19(19Y + 24X + 26)^3(3Y + X + 23)(27Y + 2X + 9)}{(X + 20)^3(X + 17)}.$$

To solve the ECDLP,

$$Q = (9, 1) = [m]P = [m](21, 24),$$

in the group generated by P, first express $(P) - (\mathcal{O})$ and $(Q) - (\mathcal{O})$ as good divisors. It is found that we can take

$$\begin{aligned} (P) - (\mathcal{O}) &= (P_1) + (P_2) - (R_1) - (R_2), \\ (Q) - (\mathcal{O}) &= (Q_1) + (Q_2) - (R_1) - (R_2), \end{aligned}$$

where $P_1 = (19 + 7\xi, 7 + 22\xi)$, $Q_1 = (11\xi, 1 + 4\xi)$ and $R_1 = (0, \xi)$ with $\xi^2 = 27$, and the points P_2, Q_2 and R_2 are the conjugates of the points P_1, Q_1 and R_1. The expressions $f_P(P)$ and $f_P(Q)$ can then be computed by computing f_P on the points P_i, Q_i and R_i. We find that $f_P(P) = 21$ and $f_P(Q) = 2$, hence $\Psi(P) = 21^4 \equiv 7 \pmod{29}$ and $\Psi(Q) = 2^4 \equiv 16 \pmod{29}$. Solving the DLP

$$7^m \equiv 16 \pmod{29}$$

it is found that $m = 5$.

V.3. The Anomalous Attack

The attack on 'anomalous' curves which was proposed by Smart [**153**], and Satoh and Araki [**136**], is explained. A similar attack has been proposed by Semaev [**143**] to solve the ECDLP in subgroups of order p, where p is the characteristic of the field of definition of the curve. In what follows use is made of the theory of elliptic curves defined over the p-adic numbers, $\mathbb{Q}_p$. For those readers unfamiliar with this area the main results are briefly summarized. For full details the reader should consult the book by Silverman [**147**].

One should think of a p-adic number as a formal base p expansion which encodes properties modulo powers of p. If n is a non-zero p-adic number it can be written in the form

$$n = p^\alpha \left(\sum_{i=0}^{\infty} n_i p^i \right),$$

where $\alpha \in \mathbb{Z}$, $n_i \in \{0, \ldots, p-1\}$ and $n_0 \neq 0$. Define $\mathrm{ord}_p(n) = \alpha$ and $|n|_p = p^{-\alpha}$. Such numbers are added and multiplied not using the power series but using the property that the result should be the correct answer if we had worked modulo any given power of p when considering the numbers as rationals modulo the given power of p (see [**151**, Chapter II]). Alternatively you could use the power series but then you need to worry about various carry operations, which is not an efficient way of proceeding.

Let E denote an elliptic curve defined over the field of p-adic numbers, $\mathbb{Q}_p$, which is assumed to have good reduction at p. The set of points of $E(\mathbb{Q}_p)$ which reduce to zero modulo p is denoted by $E_1(\mathbb{Q}_p)$ which is a group. The set of points in $E(\mathbb{Q}_p)$ which reduce modulo p to an element of $E(\mathbb{F}_p)$ is denoted by $E_0(\mathbb{Q}_p)$. In our case of E having good reduction at p we have $E(\mathbb{Q}_p) = E_0(\mathbb{Q}_p)$ but to remain consistent with the more general literature we shall still retain the notation $E_0(\mathbb{Q}_p)$. There is the exact sequence

$$0 \longrightarrow E_1(\mathbb{Q}_p) \longrightarrow E_0(\mathbb{Q}_p) \longrightarrow E(\mathbb{F}_p) \longrightarrow 0.$$

Hence multiplying an element of $E_0(\mathbb{Q}_p)$ by a multiple of the number of elements in $E(\mathbb{F}_p)$ will produce a result which lies in $E_1(\mathbb{Q}_p)$.

The group $E_1(\mathbb{Q}_p)$ is isomorphic to the group of $p\mathbb{Z}_p$-valued points of the one-parameter formal group associated to E (see [**147**, p. 175]). The isomorphism is given by

$$\vartheta_p : \begin{cases} \hat{E}(p\mathbb{Z}_p) & \longrightarrow & E_1(\mathbb{Q}_p) \\ z & \longmapsto & \begin{cases} \mathcal{O} & \text{if } z = 0, \\ \left(\frac{z}{w(z)}, \frac{-1}{w(z)}\right) & \text{otherwise, i.e. } z = -x/y, \end{cases} \end{cases}$$

where $w(z)$ is the power series in z, which is the formal power series solution to the equation

$$w = z^3 + a_1 z w + a_2 z^2 w + a_3 w^2 + a_4 z w^2 + a_6 w^3.$$

Such a solution can be computed to any desired number of terms using the standard Newton–Raphson iteration. Using the power series for $w(z)$ the Laurent series for $x(z)$, $y(z)$ and $\omega(z)$ can be computed, where $\omega(z)$ denotes the invariant differential on $\hat{E}(p\mathbb{Z}_p)$ (again see [**147**]). These Laurent series have their first few terms given by

$$\begin{aligned} x(z) &= \frac{z}{w(z)} = \frac{1}{z^2} - \frac{a_1}{z} - a_2 - a_3 z - (a_4 + a_1 a_3) z^2 - \cdots, \\ y(z) &= \frac{-1}{w(z)} = \frac{-1}{z^3} + \frac{a_1}{z^2} + \frac{a_2}{z} + a_3 + (a_4 + a_1 a_3) z + \cdots, \\ \omega(z) &= \frac{dx(z)}{2y(z) + a_1 x(z) + a_3} \\ &= \left(1 + a_1 z + (a_1^2 + a_2) z^2 + (a_1^3 + 2a_1 a_2 + a_3) z^3 + \cdots\right) dz \\ &= (1 + d_1 z + d_2 z^2 + d_3 z^3 + \cdots) dz. \end{aligned}$$

For points on $E_1(\mathbb{Q}_p)$ define the p-adic elliptic logarithm to be the group homomorphism

$$\vartheta_p : \begin{cases} E_1(\mathbb{Q}_p) & \longrightarrow \hat{\mathbb{G}}_a \\ P & \longmapsto \int \omega(z_p) = z_p + \frac{d_1 z_p^2}{2} + \frac{d_2 z_p^3}{3} + \cdots . \end{cases}$$

There is another subgroup of $E(\mathbb{Q}_p)$ which interests us, namely

$$E_2(\mathbb{Q}_p) = \{P \in E(\mathbb{Q}_p) : \mathrm{ord}_p(P_X) \leq -4\} \cup \{\mathcal{O}\},$$

which corresponds to $\hat{E}(p^2\mathbb{Z}_p)$. Here, P_X denotes the x-coordinate of the point P. This group also is involved in an exact sequence, namely

$$0 \longrightarrow E_2(\mathbb{Q}_p) \longrightarrow E_1(\mathbb{Q}_p) \longrightarrow \mathbb{F}_p^+ \longrightarrow 0,$$

which tells us that if we multiply an element in $E_1(\mathbb{Q}_p)$ by a multiple of p we will obtain an element of $E_2(\mathbb{Q}_p)$.

The crucial mathematical point which makes the following method work is that if the number of elements of $E(\mathbb{F}_p)$ is equal to the number of elements of $\mathbb{F}_p^+$ then we have the following isomorphism:

$$E_0(\mathbb{Q}_p)/E_1(\mathbb{Q}_p) \cong E_1(\mathbb{Q}_p)/E_2(\mathbb{Q}_p) \cong \mathbb{F}_p^+.$$

It will be assumed that our elliptic curve, E, is defined over a prime finite field, $\mathbb{F}_p$, and that the number of points on E is equal to p. Hence the trace of Frobenius is equal to one and the curve is said to be 'anomalous'. Consider the two points on the curve, $\overline{P}$ and $\overline{Q}$, and suppose the solution is required to the following ECDLP on $E(\mathbb{F}_p)$:

$$\overline{Q} = [m]\overline{P},$$

for some integer m. An arbitrary lift of $\overline{P}$ and $\overline{Q}$ to points, P and Q, on an elliptic curve defined over $\mathbb{Q}_p$, whose reduction is the elliptic curve $E(\mathbb{F}_p)$, is first computed. This is trivial in practice, since, as neither $\overline{P}$ nor $\overline{Q}$ is a point of order two, we can write $P = (x, y)$ where x is some lift of $\overline{P}_X$ and y is computed via Hensel's Lemma.

It follows that

$$Q - [m]P = R \in E_1(\mathbb{Q}_p).$$

Note that

$$E_0(\mathbb{Q}_p)/E_1(\mathbb{Q}_p) \cong E(\mathbb{F}_p) \text{ and } E_1(\mathbb{Q}_p)/E_2(\mathbb{Q}_p) \cong \mathbb{F}_p^+.$$

But the groups $E(\mathbb{F}_p)$ and $\mathbb{F}_p^+$ have the same order by assumption, namely p, and so

$$[p]Q - [m]([p]P) = [p]R \in E_2(\mathbb{Q}_p).$$

If the p-adic elliptic logarithm, ϑ_p, is taken of every term in the above equation we obtain

$$\vartheta_p([p]Q) - m\vartheta_p([p]P) = \vartheta_p([p]R) \equiv 0 \pmod{p^2}.$$

This is possible since for any point $P \in E(\mathbb{Q}_p)$ we have $[p]P \in E_1(\mathbb{Q}_p)$, as $p = \#E(\mathbb{F}_p)$, and the p-adic elliptic logarithm is defined on all points in

$E_1(\mathbb{Q}_p)$. Computing the p-adic elliptic logarithm is an easy matter. The value m is deduced from the equation

$$m \equiv \frac{\vartheta_p([p]Q)}{\vartheta_p([p]P)} \pmod p.$$

Clearly, on the assumption that one knows the group order, the above observation will solve the ECDLP in linear time. To see this notice that the only non-trivial steps required are the computations of $[p]P$ and $[p]Q$, both of which take $O(\log p)$ group operations on E. With probability $1/p$ the above method will fail to find the required discrete logarithm as we will obtain $\vartheta_p([p]P) \equiv 0$. However, a different curve $E(\mathbb{Q}_p)$ can then be chosen which reduces to $E(\mathbb{F}_p)$ and the method repeated.

V.3.1. Example. To explain the method a curve over a small field, namely $\mathbb{F}_{43}$, will be used. Consider the curve

$$E : Y^2 = X^3 + 39X^2 + X + 41.$$

The group $E(\mathbb{F}_{43})$ can be readily verified to have 43 elements. On this curve the ECDLP given by

$$Q = [m]P$$

is to be solved, where $P = (0, 16)$ and $Q = (42, 32)$. The following 'lifts' of these points to elements of $E(\mathbb{Q}_p)$ using Hensel's Lemma are found:

$$\begin{aligned} P &= (0, 16 + 20 \cdot 43 + O(43^3)), \\ Q &= (42, 32 + 20 \cdot 43 + +O(43^3)). \end{aligned}$$

The computation of $[43]P$ and $[43]Q$ is required and they are found to be

$$\begin{aligned} [43]P &= (38 \cdot 43^{-2} + O(43^{-1}), 41 \cdot 43^{-3} + O(43^{-2})), \\ [43]Q &= (25 \cdot 43^{-2} + O(43^{-1}), 39 \cdot 43^{-3} + O(43^{-2})). \end{aligned}$$

A simple computation reveals that

$$\begin{aligned} \vartheta_{43}([43]P) &= 19 \cdot 43 + O(43^2), \\ \vartheta_{43}([43]Q) &= 17 \cdot 43 + O(43^2), \end{aligned}$$

and so

$$m = \frac{\vartheta_{43}([43]Q)}{\vartheta_{43}([43]P)} = 19 + O(43).$$

It is concluded that m is equal to 19, which is easily verified to be correct.

V.4. Baby Step/Giant Step

We describe the *baby step/giant step* (BSGS) method for a general finite abelian group, G, with n elements. By the Pohlig–Hellman simplification it can be assumed that n is prime; however, this fact is not used in the description below. Let $P, Q \in G$ with

$$Q = [m]P.$$

The value of m is sought. By simple Euclidean division it is known that m can be written as

$$m = \lceil \sqrt{n} \rceil a + b$$

with $0 \leq a, b < \lceil \sqrt{n} \rceil$. The only problem is that the values of a and b are not known. The equation is rewritten to look for a solution in terms of a and b of

$$(Q - [b]P) = [a]([\lceil \sqrt{n} \rceil]P).$$

This may seem like an added complication but it allows us to perform a standard space/time trade off. This idea is the BSGS method and is due to Shanks in the context of factoring algorithms and class group computations.

A table of 'baby steps' is first computed. This is a table of all values of

$$R_b = Q - [b]P$$

where b ranges between 0 and $\lceil \sqrt{n} \rceil - 1$. This table should be sorted on the R_b and stored in memory so that it can be efficiently searched by using a binary search method. This adds the complication that one needs to have a way of comparing elements of the group. In practice this is no problem since if each element has a unique representation then the bit representation of the element in the computer will be sufficient as a key.

After having computed the 'baby steps' the 'giant steps' are computed:

$$S_a = [a]([\lceil \sqrt{n} \rceil]P).$$

On each computation of a giant step it is seen whether S_a occurs in the table. If it does the values of a and b are recovered. By an earlier comment, this method must terminate before a reaches the value of $\lceil \sqrt{n} \rceil$.

The complexity of the method is roughly $O(\sqrt{n})$ as this much time is required to compute the baby steps and a maximum of this amount of time to compute the giant steps. In this complexity estimate we have ignored the time needed to perform the table look up. However, the main problem with the method is that it requires the storage of $O(\sqrt{n})$ group elements.

The rho and lambda methods, discussed in more detail in the next subsection, also have complexity $O(\sqrt{n})$. The time to sort and search the look-up table in the BSGS method can be eliminated if a hash table is used instead. In this case, the constant multiplying $\sqrt{n}$ in the asymptotic estimate can be made $\frac{4}{3}$. Pollard's rho method has a slightly better constant, roughly $\frac{5}{4}$. However, this is only an *expected* running time, given the randomized nature of the method. The lambda method, on the other hand, has a constant of 2, again applied to expected running time. The advantage of the rho and lambda methods is that their storage requirements can be made arbitrarily small.

V.4.1. Example. As promised earlier, the elliptic curve

$$E : Y^2 = X^3 + 71X + 602$$

over $\mathbb{F}_{1009}$ is again considered and the ECDLP is to determine m_0 in

$$Q' = (592, 97) = [m_0](32, 737) = [m_0]P',$$

in the subgroup of order 53 generated by $P' = (32, 737)$. Since $\lceil\sqrt{53}\,\rceil = 8$ eight baby steps are required. So we compute

b	$R_b = Q' - [b]P'$
0	(592, 97)
1	(728, 450)
2	(537, 344)
3	(996, 154)
4	(817, 136)
5	(365, 715)
6	(627, 606)
7	(150, 413)

The giant steps $[a]([8]P')$ are computed as

a	$[a]([8]P')$
1	(996, 855)
2	(200, 652)
3	(378, 304)
4	(609, 357)
5	(304, 583)
6	(592, 97)

It is seen that a match is obtained with $a = 6$ and $b = 0$, which implies that the solution to the DLP is given by

$$m_0 = a8 + b = 6 \cdot 8 + 0 \equiv 48 \pmod{53}.$$

The solution to the original problem, posed in Section V.1, is given by the unique positive integer less than 530 which is congruent to 1, 4 and 48 modulo 2, 5 and 53 respectively. So the solution is 419, which is easily checked to be correct.

Notice that in this example, the giant steps up to $a = 6$ need not have been computed. We could have halted at $a = 1$ and noticed that

$$[8]P' = -R_3 = -Q' + [3]P'.$$

Hence

$$[m_0]P' = Q' = [3]P' - [8]P' = [48]P',$$

which leads again to $m_0 \equiv 48 \pmod{53}$.

V.5. Methods based on Random Walks

Pollard [**125**] gives a number of methods to solve the discrete logarithm problem in a variety of groups. The rho method uses a single random walk and waits for a cycle to occur. By using a space-efficient method to detect the cycle, the discrete logarithm can be found. The wait for the cycle means that the single random walk can be thought of as tracing out the greek letter rho, ρ.

In Pollard's lambda method (often called the method of tame and wild kangaroos), two random walks are used, one by a tame kangaroo who jumps off into the wild, digs a hole and waits for the wild kangaroo to fall into it. The two paths form the shape of the greek letter lambda, λ. The lambda method is suited to finding discrete logarithms which are known to lie in a short interval.

There is a parallel version of the rho method, which uses many random walks. However, despite the method's name, the 'paths' do not now look like a rho, since instead one looks for two paths that intersect. The method described in this section is what is usually referred to as the parallel rho method.

The following intuitive explanation uses the analogy of jumping animals, since we have found this to be useful when explaining the method in lectures. However, these are not the kangaroos of Pollard's method, since Pollard's kangaroos perform better controlled jumps. We shall call our jumping animals 'snarks', since they jump around in a rather uncontrolled manner.

To simplify the matter we take two snarks. Eventually we shall use a larger number of snarks. The two snarks are given a spade and told that they should dig a hole every ten or so jumps. Where each snark jumps next depends on the position they are currently at, hence when one snark meets the path of the other (or itself) it will follow the original path along until it falls into one of the holes that have been dug.

If both snarks are jumping around a field of finite size then eventually the path of the one will intersect the path of the other. This may seem a doubtful strategy but the philosophy can be easily turned into a general purpose method for solving discrete logarithms.

We explain the method for a general finite abelian group, G, of order n. Let $P, Q \in G$ with

$$Q = [m]P,$$

and again we wish to find m. For our method the following two functions will be needed:

$$f : G \longrightarrow \{1, \ldots, s\}$$

for some positive integer s to be determined, and

$$H : G \longrightarrow \mathbb{Z},$$

a hash function from the group G to the integers. It will be assumed the map f provides an 'equidistribution' function in the sense that

$$\sum_{i=1}^{s} \left| |\{g \in G : f(g) = i\}| - \frac{n}{s} \right| = O(\sqrt{n}).$$

It will be assumed in this application that the only collisions the function $H(g)$ has are between g and $-g$.

A set of multipliers, M_i, for $i = 1, \ldots, s$, of the form
$$M_i = [a_i]P + [b_i]Q,$$
for random $a_i, b_i \in \mathbb{Z}$ is fixed and the following function is defined:
$$F : \begin{cases} G & \longrightarrow & G \\ g & \longmapsto & g + M_{f(g)} \end{cases}$$
This function, given any starting point, g_0, in G defines a random walk, $g_k = F(g_{k-1})$. Such a walk is efficient to implement and has important statistical properties. Practical experiment [**157**] has shown that choosing $s \approx 20$ gives the correct balance between statistical behaviour and efficiency, for group orders of a reasonably large size. These properties have led this walk to be considered for various algorithms in finite abelian groups (see [**79**], [**137**] and [**158**]).

The method goes as follows: Take two elements (snarks) in the group G of the form
$$\begin{aligned} g_0 &= [x_0]P + [x_0']Q, \\ h_0 &= [y_0]P + [y_0']Q. \end{aligned}$$
Apply the random walk defined above to compute
$$\begin{aligned} g_k &= [x_k]P + [x_k']Q, \\ h_k &= [y_k]P + [y_k']Q. \end{aligned}$$
After about $O(\sqrt{n\pi/2})$ iterations we will find the paths have crossed so that
$$[x_k]P + [x_k']Q = g_k = h_l = [y_l]P + [y_l']Q$$
for some k and l. Thus
$$[x_k - y_l]P = [y_l' - x_k']Q = [y_l' - x_k'][m]P,$$
and, given the group order n, the DLP can be solved, unless we are very unfortunate and $\gcd(y_l' - x_k', n) \neq 1$.

The main point is that once the two snarks paths have crossed they will travel along on the same road. This is because both are following the same random walk algorithm. A diagram of their paths will look like the Greek letter lambda, λ.

Since both snarks follow the same random walk algorithm, it is not necessary to store a large number of points, g_i, and see whether each point h_i is the same as any of the g_i, in order to detect collisions. Instead, only those elements g_i (or h_j) for which $H(g_i)$ satisfies some certain arithmetic property, such as its last k bits being all zero, are stored. Such an element will be called 'distinguished' [**128**]. If this is the case then the expected total storage, $O(\sqrt{n\pi/2}/2^k)$, can be made as small as can be coped with. This is achieved at the expense of each snark having to jump for, on average, an extra 2^k steps.

Notice that a solution is also obtained if $g_k = g_l$ for some k and l with $k \neq l$. Hence only one snark could be used if we wanted (this is usually called the *rho* method as the path of the snark will eventually form a shape like the Greek letter rho, ρ).

As van Oorschot and Wiener [**116**] point out this method can be trivially parallelized, by having n rather than two snarks. However, unlike the standard parallelization of the *rho* method, the parallelization of van Oorschot and Wiener provides a linear speed up. So n snarks will solve the DLP twice as fast as $n/2$ snarks.

In practice a set of client programs perform the random walks of the snarks, with say one snark per client. They then pass any distinguished points they find back to a server who collects the distinguished points in a database and searches for matches.

Some implementation details of this parallelization for elliptic curves are discussed. Each client program can perform a number of random walks in parallel as this allows us to perform an efficient 'parallel' inversion [**106**]. Hence each client program is actually computing a set of random walks and not just one. For the hash function the value of the x-coordinate on the curve can be used, whilst the distinguished points will be those with a certain number of least significant bits of the x-coordinate equal to zero.

When a report is received which has an identical hash value with one already found, the DLP can be solved. Actually two problems are solved, as the function $H(g)$ discards information about the y-coordinate. Once solved the answers can be checked as to which solution is the correct one.

V.5.1. Example. The elliptic curve

$$E : Y^2 = X^3 + 71X + 602$$

over $\mathbb{F}_{1009}$ is again considered and a solution to the ECDLP

$$Q' = (592, 97) = [m_0](32, 737) = [m_0]P'$$

in the subgroup of order 53 generated by $P' = (32, 737)$, sought. It was shown earlier using the BSGS method that this had the solution 48. This result will be established again using the method of random walks.

Some choices are first made. As the example is small, we choose $s = 3$ and define

$$f : \begin{cases} E(\mathbb{F}_{1009}) & \longrightarrow & \{1, 2, 3\} \\ (x, y) & \longmapsto & (x \pmod 3) + 1. \end{cases}$$

The multipliers are chosen to be

$$\begin{aligned} M_1 &= [2]P' + [0]Q' = (8, 623), \\ M_2 &= [1]P' + [1]Q' = (654, 118), \\ M_3 &= [3]P' + [4]Q' = (555, 82). \end{aligned}$$

Since the group order is very small assume that every point is distinguished. The two snarks are called g and h and their positions at time t will be denoted by g_t and h_t. Initially set $g_0 = P'$ and $h_0 = Q'$ and let the snarks pursue their 'random' walk:

t	g_t	h_t
0	$[1]P' + [0]Q' = (32, 737)$	$[0]P' + [1]Q' = (592, 97)$
1	$[4]P' + [4]Q' = (200, 357)$	$[1]P' + [2]Q' = (817, 136)$
2	$[7]P' + [8]Q' = (759, 545)$	$[2]P' + [3]Q' = (304, 583)$
3	$[9]P' + [8]Q' = (241, 691)$	$[3]P' + [4]Q' = (555, 82)$
4	$[10]P' + [9]Q' = (711, 716)$	$[5]P' + [4]Q' = (809, 516)$
5	$[12]P' + [9]Q' = (759, 545)$	—

So the g-snark has crossed its own path, as $g_2 = g_5$. Its path looks like a rho.

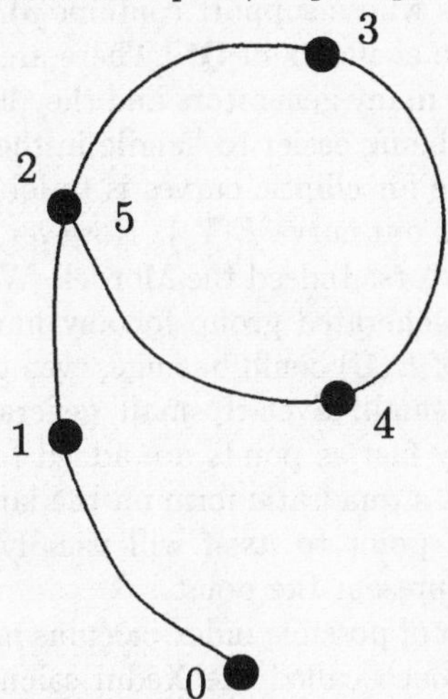

The ECDLP is solved using the resulting equation

$$[7]P' + [8]Q' = g_2 = g_7 = [12]P' + [9]Q'$$

and so

$$[5]P' = -Q' = -[m_0]P'.$$

Hence $m \equiv -5 \equiv 48 \pmod{53}$, as required.

V.6. Index Calculus Methods

One reason for proposing elliptic curves in cryptography is that there appears to be no analogue of the index calculus methods which are available for schemes using the multiplicative groups of finite fields.

Index calculus methods usually use a set of elements, called the *factor base*. On this factor base a set of relations are found. Once a full lattice of relations is determined one can solve virtually any DLP in a straightforward manner. At least one can do this as soon as the elements defining the DLP have been expressed in terms of the factor base.

There are two 'philosophical' ways of designing index calculus type methods for elliptic curves. Both however lead to significant problems.

The first uses the fact that the group of points on an elliptic curve is in fact the class group of a function field. Ideas like the function field sieve [**1**] can be used to obtain an index calculus method. In Chapter X it will be seen that for hyperelliptic curves of large genus this does indeed give a sub-exponential method. However, for elliptic curves the method is very ineffective as the factor base needs to consist of all the points on the elliptic curve.

The second 'philosophy' is that index calculus methods for the groups $\mathbb{F}_p^*$ make use of the fact that $\mathbb{F}_p^*$ 'is' the reduction modulo p of the group $\mathbb{Q}^*$ (apart from the elements whose support contains p). The factor base is then chosen to be small prime elements in $\mathbb{Q}^*$. There are a lot of these to choose from as $\mathbb{Q}^*$ has infinitely many generators and they have an obvious ordering, with smaller generators being easier to handle in the computer.

The obvious analogue for elliptic curves is to look at curves $E(\mathbb{Q})$ whose reduction modulo p gives our curve $E(\mathbb{F}_p)$. However, $E(\mathbb{Q})$ does not have an infinite number of generators. Indeed the Mordell–Weil Theorem [**147**] states that $E(K)$ is a finitely generated group for any number field K. Not only that but the generators of $E(\mathbb{Q})$ could be huge, even though the coefficients of E can be comparatively small. Even if 'small' generators exist the size of the rational points grow very fast as points are added together. This is because the Neron–Tate height is a quadratic form on the lattice of the Mordell-Weil group. Hence adding a point to itself will usually double the size of the coordinates needed to represent the point.

For a fuller discussion of possible index calculus methods see [**149**]. There is another possible approach called the Xedni calculus (see [**150**]; Xedni is indeX backwards). This method uses a number of deep ideas from the theory of elliptic curves over the rationals such as the Birch–Swinnerton-Dyer conjecture, see [**12**] and [**13**]. However, this method appears, at present, unlikely to yield a practical solution for the ECDLP – see [**55**] for a detailed discussion.

V.7. Summary

To summarize, let E be an elliptic curve over $\mathbb{F}_q$, with group order $n = \#E(\mathbb{F}_q)$. For E to be used in a cryptosystem, we require the following properties:

1. The group should have a subgroup of large prime order r, where the meaning of 'large' is determined by the desired cryptographic strength, based on the best implementations of square root attacks using current software and hardware. This is often taken to mean a prime of more than 160 bits, which compares to the security of about 1000 bits of key

length in conventional public key systems as implied by current algorithmic knowledge (see Section I.3). From the point of view of efficient cryptosystem implementation, the ratio of cryptographic strength to computational cost is maximized when $\log r$ is close to $\log n$ (this is discussed in more detail in Section VI.5).

2. The curve should not be anomalous, i.e. $q = n = p$, a prime. These are the curves of trace one over $\mathbb{F}_p$.
3. The smallest value of l such that $q^l \equiv 1 \pmod{n}$ should be large. This removes curves of trace zero and two over $\mathbb{F}_p$ immediately, as well as the other supersingular curves.

Note that all of the above conditions are very easy to check as soon as a curve and its group order have been computed.

Property 3 (sometimes referred to as the *MOV condition* [**P1363**]) aims at preventing the possibility of embedding the ECDLP in the multiplicative group of a finite field with an easier DLP, as done in the MOV attack. To quantify the meaning of 'large l' in this context, we recall the complexity estimates for the DLP and the general ECDLP from Section I.3. The goal is to have

$$C_{\text{EC}}(k) \approx C_{\text{CONV}}(lk),$$

where $k = \lceil \log_2 q \rceil$, $C_{\text{EC}}(k)$ denotes the complexity of the ECDLP for curves over $\mathbb{F}_q$, and $C_{\text{CONV}}(lk)$ denotes the complexity of the DLP on $\mathbb{F}_{q^l}$. Using the estimates for C_{EC} and C_{CONV} from Section I.3, it is readily verified that the goal is achieved when

$$l = O\left(\frac{k^2}{(\log k)^2}\right).$$

CHAPTER VI

Determining the Group Order

The problem of determining the order of the group of rational points on an elliptic curve over a finite field – the *point counting problem* – is of critical importance in applications such as primality proving and cryptography. As seen in the summary section of Chapter V, for cryptographic applications, we require the curve to be non-supersingular, and the group order to be divisible by a large prime factor, which in practice may be several hundred bits long (160 bits is sometimes considered a minimal requirement). Therefore, the problem is difficult, and it requires innovative solutions that are both mathematically challenging and computationally effective.

The point counting problem is introduced in this chapter, where general methods for finite groups, as well as some 'easier' cases of elliptic curve groups, are discussed. More advanced methods applicable to broader classes of curves are discussed in Chapters VII and VIII.

VI.1. Main Approaches

Three main techniques are presently used to determine elliptic curves suitable for cryptography:

- Generate random curves and compute their group orders, until an appropriate one is found.
- Generate curves with given group order using the theory of complex multiplication (CM). Such curves are usually called CM-curves (which is somewhat misleading, as all curves over a finite field have CM through the Frobenius map).
- Use the group of $\mathbb{F}_{q^n}$-rational points, $E(\mathbb{F}_{q^n})$, of a curve E defined over $\mathbb{F}_q$, for q relatively small. Taking a view centered on the field over which the rational points of interest are defined, the curve in this case is often referred to as a *subfield curve*, or a curve of *Koblitz type.*

One approach to computing the order of a general finite abelian group is to use a generalization of either the BSGS method or the methods based on random walks (the rho, lambda and kangaroo methods) discussed in the previous chapter. To this end, we compute the order of a randomly chosen element $g \in G$, i.e. determine n so that

$$g^n = e$$

where e is the group identity. Considering the orders obtained for several randomly chosen group elements will give a possible value for the group order. To obtain more certain information, subtler methods are required, (see [**21**] and [**158**]).

By Hasse's Theorem it is known that the number of rational points on an elliptic curve over $\mathbb{F}_q$ satisfies

$$|q+1-\#E(\mathbb{F}_q)| \leq 2\sqrt{q},$$

and the group of points is the product of at most two cyclic subgroups.

A naive way of counting rational points on curves over small finite fields of odd characteristic p, with curve equation given by

$$Y^2 = X^3 + aX + b,$$

is to evaluate the sum

$$p+1+\sum_{x=0}^{p-1}\left(\frac{x^3+ax+b}{p}\right)$$

where $\left(\frac{\cdot}{p}\right)$ is the Legendre symbol. This is reasonably fast for small values of p but soon becomes unwieldy for large values. Cohen [**29**] suggests using this method for $p < 10000$. He also notes that combining Shanks's BSGS method with Hasse's Theorem will give an $O(p^{1/4+\epsilon})$ method, where ϵ is a positive constant that can be made arbitrarily small. This is the method of Shanks and Mestre, which is claimed to perform better than the Legendre symbol method for $p > 457$ [**29**]. We discuss the Shanks–Mestre method in Section VI.3. For large finite fields, i.e., the typical situation when looking for elliptic curves for use in cryptography, a better method is needed.

The method currently believed to yield curves least amenable to attack consists of choosing a large finite field and then selecting elliptic curves over that field at random until one is found whose group of rational points satisfies the group order constraints. This procedure is outlined in Section VI.5, where the probability of success in each trial is estimated. The procedure requires the ability to determine the number of rational points for an arbitrary curve over a large finite field. This task is computationally challenging but feasible for field sizes of practical interest, and is the main subject of Chapter VII.

Another way of proceeding is to decide on a prime base field of large order and then use the theory of curves with CM to produce curves with a cyclic subgroup of large prime order [**73**]. Again, this is feasible but some involved computations are still required, e.g., computing roots of large degree polynomials over large finite fields. Nevertheless, the CM method is considered less computationally taxing than general point counting. This method is the subject of Chapter VIII.

When general point counting on random curves, or the CM method, is deemed too complex, a reasonable compromise can be found in the use of subfield curves. This method, described in Section VI.4 below, enjoys popularity due to the ease by which appropriate curves over very large fields can be produced. However, the family of such curves is rather small compared to the family of general curves over $\mathbb{F}_{q^n}$, offering less choice in the design of a cryptosystem, and making the curves somewhat 'special'. This has lead some researchers in the cryptographic community to express concerns that such curves might be weaker, in the sense that their associated DLP might be easier, than those generated in a random way. The same concern about the curves being 'special' has been expressed about curves generated with the CM method. As of the writing of this book, however, no DLP algorithm that actually exposes any weaknesses has been found for curves generated using either the CM or the subfield method.

A common feature of some of the point counting algorithms presented is that they produce one or more 'candidate' values m for the group order, and need to check whether m is indeed the order. In the next section, we discuss how this is done.

VI.2. Checking the Group Order

Given an elliptic curve E defined over $\mathbb{F}_q$, we wish to determine whether a given integer m produced by a point counting algorithm is the order of $E(\mathbb{F}_q)$. The first obvious test is to ascertain that m is within the Hasse interval

$$q + 1 - 2\sqrt{q} \leq m \leq q + 1 + 2\sqrt{q}.$$

Once this is established, a point P in $E(\mathbb{F}_q)$ is selected at random (e.g., using Algorithm III.1), and the condition

$$[m]P = \mathcal{O}$$

is checked. Clearly, if the condition does not hold, m is not the group order. If the condition holds, there is a high likelihood of m being the group order, the exact probability depending on the factorization of m. The probability can be increased by drawing and checking further random points. One possible approach is to keep all candidates m that pass a random point test, and at the end of the algorithm draw and check random points until only one candidate survives. This approach succeeds if the point counting algorithm is guaranteed to produce the true group order as one of the candidates, which is the case for the algorithms described in this and the next two chapters.

For the group orders of interest in cryptography, the situation is simpler. In this case, we are only interested in group orders m of the form

$$m = s \cdot r,$$

where s is a small positive integer, and r is prime. First, it is easy to check that a candidate m is indeed of this form. For values of s used in practice, this can be done by trial division and primality testing. If m is not of the desired form, then it is discarded, even though it might be the true group order. If no candidates survive at the end of the point counting algorithm, the curve is deemed inappropriate for cryptographic applications, and another one is tried. When m is of the right form, a random point P is checked. If $[m]P = \mathcal{O}$, the multiple $[s]P$ is checked. If $[s]P = \mathcal{O}$ (this has extremely low probability), then P is discarded and a new random point is checked. Otherwise, r divides the order of P. If $r > 4\sqrt{q}$, this condition guarantees that m is the group order, as there can be no other multiple of r in the Hasse interval. The condition on r is amply met in practical applications, since $s \ll r$ (the choice of s is discussed in more quantitative terms in Section VI.5).

Several related results exist in the literature. For example, a theorem of Mestre, quoted in [**142**], shows that for p a prime greater than 461, and a curve defined over $\mathbb{F}_p$, either the curve or its twist will always admit a point of order greater than $4\sqrt{p}$. Schoof [**142**] extends this result by showing that for $p > 229$ either the curve or its twist admits an $\mathbb{F}_p$-rational point P with the property that the only integer $m \in (p+1-2\sqrt{p}, p+1+2\sqrt{p})$ for which $[m]P = \mathcal{O}$ is the group order.

VI.3. The Method of Shanks and Mestre

We give a brief description of the $O(q^{1/4+\epsilon})$ BSGS-based algorithm for determining the group order of an elliptic curve. If the group order is expressed as $\#E(\mathbb{F}_q) = q+1-t$, $|t| \leq 2\sqrt{q}$, the uncertainty is of order $4\sqrt{q}$ and an algorithm of order $q^{1/4+\epsilon}$ is not surprising. The algorithm uses an idea of Mestre and is referred to as the Shanks–Mestre algorithm [**29**]. The discussion below is adapted from [**40**].

Determine a random point P on $E(\mathbb{F}_q)$. It is assumed the order of P is greater than $4\sqrt{q}$ (see the discussion in Section VI.2). Let $Q = [q+1]P$ and $Q_1 = Q + [\lfloor 2\sqrt{q} \rfloor]P$ and let $t' = t + \lfloor 2\sqrt{q} \rfloor \in [0, 4\sqrt{q}]$. Let $m = \lceil 2q^{1/4} \rceil$ and observe that $t' = im + j$ for some positive integers $i, j < m$. Compute the baby steps $[j]P$, $j = 0, 1, 2, \ldots, m-1$, and store in some convenient manner. For i from 0 to $m-1$ compute the giant steps $Q_1 - [i]([m]P)$ and check to see which one is in the table. If the ith giant step is equal to the jth baby step then $t' = im + j$ and t is obtained.

The algorithm takes on the order of $q^{1/4+\epsilon}$ in both time and space, where ϵ is a positive constant that can be made arbitrarily small. As usual, using the techniques of Pollard, the space requirements can be made arbitrarily small, without sacrificing the asymptotic expected running time of the algorithm. For large q, however, the above algorithm quickly becomes impractical.

VI.4. Subfield Curves

Let E be a curve defined over $\mathbb{F}_q$, and write $N_n = \#E(\mathbb{F}_{q^n})$ for $n \geq 1$. Define the series

$$Z(E;T) = \exp\left(\sum_{n\geq 1} \frac{N_n}{n} T^n\right)$$

for an indeterminate T. This is referred to as the *zeta function* [**66**] of the curve E over $\mathbb{F}_q$. The following theorem, due to Hasse, shows that this function has a very simple form that allows all the values of N_n, $n > 1$, to be obtained from knowledge of N_1, the number of rational points over the ground field, $\mathbb{F}_q$. Owing to results by Weil (see, e.g.,[**147**]), the theorem can actually be extended to curves of genus higher than one, and this will be briefly discussed in Chapter X.

THEOREM VI.1 (see, e.g., [**66**]). *Let E be an elliptic curve over $\mathbb{F}_q$, and let c_1 denote its trace of Frobenius at q, i.e., $N_1 = q + 1 - c_1$. The zeta function of E over $\mathbb{F}_q$ has the form*

$$Z(E;T) = \frac{P(T)}{(1-T)(1-qT)},$$

where

$$P(T) = 1 - c_1 T + qT^2 = (1-\alpha T)(1-\overline{\alpha}T).$$

The discriminant of $P(T)$ is non-positive, and the magnitude of α is $\sqrt{q}$.

Notice that Theorem III.3 in Chapter III follows immediately from Theorem VI.1. Also, it follows by straightforward series manipulations and partial fraction expansions that, for $n \geq 1$,

$$N_n = \#E(\mathbb{F}_{q^n}) = q^n + 1 - \alpha^n - \overline{\alpha}^n = |1 - \alpha^n|^2. \qquad \text{(VI.1)}$$

Clearly, this equation provides an efficient computational procedure for $\#E(\mathbb{F}_{q^n})$, since α is a quadratic imaginary integer, and α^n can be computed using a binary exponentiation method (see Chapter IV), involving only operations on integers. An alternative formulation, also leading to an efficient computation, is given in the following corollary, which follows immediately from Equation (VI.1).

COROLLARY VI.2. *Let $\mathbb{F}_q$, E and c_1 be defined as in Theorem VI.1. Write $\#E(\mathbb{F}_{q^n}) = q^n + 1 - c_n$, for $n \geq 1$. Then,*

$$c_n = c_1 c_{n-1} - q c_{n-2},$$

where $c_0 = 2$.

For commonly used ranges of values of q and n, n has to be prime to allow for a large enough prime divisor of $\#E(\mathbb{F}_{q^n})$. For, if n factors non-trivially as $n = n_1 n_2$, then both $\#E(\mathbb{F}_{q^{n_1}})$ and $\#E(\mathbb{F}_{q^{n_2}})$ divide $\#E(\mathbb{F}_{q^n})$, limiting the range of the largest prime divisor of $\#E(\mathbb{F}_{q^n})$.

It is common to consider subfield curves in characteristic two, to take advantage of the efficient arithmetic in such fields. However, the trouble with restricting attention to characteristic two is that there are not many curves defined over small finite fields with the required subgroup of large prime order over the extension field. One possible way around this is to use subfield curves over fields of odd characteristic (see [**152**]).

As an example in characteristic two consider an elliptic curve, E, defined over the finite field $\mathbb{F}_4$, with equation

$$E : Y^2 + XY = X^3 + \theta + 1.$$

Here, $\theta^2 + \theta + 1 = 0$ and $\mathbb{F}_4 = \mathbb{F}_2[\theta]$. It is verified by direct inspection that $\#E(\mathbb{F}_4) = 4$, and thus, the trace of Frobenius at $q = 4$ is $c_1 = 1$. Using Equation (VI.1) or Corollary VI.2, it is seen that the group order of the rational points over $\mathbb{F}_{2^{158}} = \mathbb{F}_{4^{79}}$ is equal to

$$2^2 \cdot 91343852333181432387730573045979447452365303319,$$

this last element in the factorization being a 156-bit prime number. Note that although curves of trace one are anomalous this does not mean the attack on anomalous curves applies in this context. The above curve is anomalous over $\mathbb{F}_4$ but not over $\mathbb{F}_{4^{79}}$. All that the attack of Semaev, Smart, Satoh and Araki will give is the solution of the DLP in the subgroup of order four, which is not really a hard problem.

VI.5. Searching for Good Curves

The preferred method for generating a 'good' curve suitable for cryptographic applications is based on selecting curves at random, and determining group orders until a curve satisfying the desired conditions is found. The method is outlined in Algorithm VI.1 below.

ALGORITHM VI.1: **Generating a 'Good' Elliptic Curve.**

INPUT: A large finite field $\mathbb{F}_q$, a small positive integer $\bar{s}$.
OUTPUT: An elliptic curve E over $\mathbb{F}_q$ such that $E(\mathbb{F}_q) = s \cdot r$, $s \leq \bar{s}$, r prime.

1. Draw E at random, with coefficients in $\mathbb{F}_q$.
2. Determine $\#E(\mathbb{F}_q)$.
3. Check the MOV and anomalous conditions (see Chapter V). If any of these fail, go to Step 1.
4. Attempt to factor $\#E(\mathbb{F}_q)$ in 'reasonable' time. If the attempt fails, go to Step 1.
5. If $\#E(\mathbb{F}_q) = s \cdot r$, $s \leq \bar{s}$, r prime, return E. Else, go to Step 1.

The main step of this procedure, Step 2, is the subject of Chapter VII. Notice that the factorization attempt in Step 4 is not difficult. In fact, for values of $\bar{s}$ used in practice, the factorization can be carried out by trial division of possible factors up to $\bar{s}$, together with primality testing. If the factorization fails, the order $\#E(\mathbb{F}_q)$ is not of the desired form, and the curve is discarded.

We now estimate the probability of success in one iteration of Algorithm VI.1. First, the term 'large', used when referring to the prime r (or equivalently, 'small' when referring to $\overline{s}$), is quantified. For the sake of concreteness, $\mathbb{F}_q$ is assumed to be of characteristic two, i.e., $q = 2^n$. Similar considerations apply to the case where q is odd. Assuming that the ECDLP is indeed of exponential complexity, $\log_2 r$ is a good measure of the 'number of bits of security' in the cryptosystem, in that a search exponential in this measure is needed to break the system. On the other hand, the 'key size' is n, the size of a field element, and the complexity of the operations required to implement the cryptosystem grows polynomially with n. Therefore, to obtain the strongest possible system for the computational investment, we would like $\log_2 r$ to be as large as possible. Recall that by Hasse's Theorem, $\log_2 r$ is roughly bounded above by n. Define the *loss* of the cryptosystem as

$$\epsilon = 1 - \frac{\log_2 r}{n}.$$

For example, an elliptic curve cryptosystem over $F_{2^{200}}$ using a cyclic prime order subgroup of order near 2^{190} has a loss of five percent.

For an integer s, let H_s denote the set of multiples of s in the interval $[q+1-2\sqrt{q},\, q+1+2\sqrt{q}]$, $q = 2^n$, and and let $H_s/s = \{\, i : i \cdot s \in H_s \,\}$.

Following Koblitz [**64**], to estimate the probability of drawing a random curve with given loss ϵ, two assumptions are made: (i) the order $\#E(\mathbb{F}_q)$ is uniformly distributed in H_2, and (ii) for small s, the distribution of primes among integers in H_s/s is similar to the distribution of primes among arbitrary integers of the same order of magnitude as q/s. By the Prime Number Theorem, the density of primes in H_s/s is thus assumed to be roughly $1/\log(q/s)$. See the discussion on these assumptions in [**64**] and [**89**].

Let $S = \lfloor 2^{n\epsilon - 1} \rfloor$. Then, under the above assumptions, and using well known properties of the harmonic series, the probability of a curve of loss ϵ or less is estimated by

$$\sum_{j=1}^{S} \frac{1}{j \log(q/2j)} \geq \frac{1}{\log q} \sum_{j=1}^{S} \frac{1}{j} = \frac{1}{\log q}(\log S + O(1)) = \epsilon + o(1).$$

For example, for the target loss of five percent in the example above, we expect to have to determine the order $\#E(\mathbb{F}_q)$ for 20 random curves before a 'good' one is found. In fact, since each point counting computation yields the orders of a curve and its twist, the point counting procedure will need to be run only ten times on the average.

The probability estimates above (and the underlying assumptions) are borne out by experimental data gathered at Hewlett–Packard Laboratories for a large number of elliptic curves over fields of the sizes used in real systems. Group orders were determined using some of the point counting techniques described in Chapter VII, and a sample of the curves obtained is presented in Appendix A.

CHAPTER VII

Schoof's Algorithm and Extensions

As discussed in Chapter VI, the preferred method for generating curves suitable for cryptographic applications depends on the ability to solve the *point counting problem* for arbitrary elliptic curves over large finite fields, namely, the exact determination of the number of rational points on such curves. In this chapter, approaches to the problem that have met with success are outlined. The two main cases of interest will be those fields of characteristic two and fields of large prime order p. While much of the theory for point counting on elliptic curves is quite general, the techniques that have developed for the two cases have differences. The two cases will be discussed separately in later sections, although the discussion will be for the general case as much as possible. There is considerable overlap in the basic theory.

Recall that for fields of characteristic two, we are only interested in non-supersingular curves, and it is sufficient to consider curve equations of the form

$$Y^2 + XY = X^3 + a_6, \quad a_6 \in \mathbb{F}_{2^n}^*,$$

as the group order of a twist is easily derived. For odd characteristic greater than three, the curves of interest have equations of the form

$$Y^2 = X^3 + aX + b, \quad a, \ b \in \mathbb{F}_q.$$

VII.1. Schoof's Algorithm

The genesis of the efficient general point counting algorithms lies in the work of Schoof [**141**]. In a dramatic improvement, Schoof dropped the complexity of the methods for point counting from $O(q^{1/4+\epsilon})$ for every positive ϵ, which results from the complexity of the BSGS algorithm applied to this case, to $O(\log^8 q)$ [**142**]. Schoof's algorithm, described in this section, forms the basis of all current efficient schemes for point counting.

By Hasse's Theorem, $\#E(\mathbb{F}_q) = q + 1 - t$ where $|t| \le 2\sqrt{q}$. The heart of Schoof's algorithm is the determination of t modulo primes ℓ for $\ell \le \ell_{\max}$ where $\ell_{\max}$ is the smallest prime such that

$$\prod_{\substack{\ell \text{ prime} \\ 2 \le \ell \le \ell_{\max}}} \ell > 4\sqrt{q}.$$

It then follows from the Chinese Remainder Theorem (CRT) that the value of t can be recovered uniquely and the group order obtained. From the

Prime Number Theorem it readily follows that the number of primes needed is $O(\log q/\log\log q)$ and that the size of the $\ell_{\max} = O(\log q)$.

A brief overview of the basic Schoof algorithm is first given, followed by a more detailed discussion on the actual computations.

Notice that one can easily determine $t \pmod{\ell}$ for $\ell = 2$, for either of the two types of field considered. For the case of odd characteristic, we have $t \equiv \#E(\mathbb{F}_q) \pmod 2$, and we saw in Section III.3.1 that $\#E(\mathbb{F}_q) \equiv 1 \pmod 2$ if and only if $X^3 + aX + b$ is irreducible over $\mathbb{F}_q$. The latter condition is equivalent to $\gcd(X^3 + aX + b, X^q - X) = 1$. For characteristic two, since the curve is non-supersingular, we have $t \equiv 1 \pmod 2$.

We now consider primes $\ell > 2$. Recall from Chapter III that the Frobenius endomorphism φ of the curve is the map given by

$$\varphi : \begin{cases} E(\overline{\mathbb{F}}_q) & \longrightarrow & E(\overline{\mathbb{F}}_q) \\ (x,y) & \longmapsto & (x^q, y^q), \\ \mathcal{O} & \longmapsto & \mathcal{O}, \end{cases}$$

and for any $P \in E(\overline{\mathbb{F}}_q)$ it satisfies the equation

$$\varphi^2(P) - [t]\varphi(P) + [q]P = \mathcal{O}. \tag{VII.1}$$

We consider this equation for points in $E[\ell]^* = E[\ell]\backslash\{\mathcal{O}\}$. Let $q_\ell \equiv q \pmod{\ell}$, and $t_\ell \equiv t \pmod{\ell}$, where the least non-negative representative of the congruence class is taken as q_ℓ and t_ℓ. If a value of $\tau \in \{0, 1, \ldots, \ell-1\}$ is found such that for a point $P = (x, y) \in E[\ell]^*$ we have

$$(x^{q^2}, y^{q^2}) + [q_\ell](x, y) = [\tau](x^q, y^q) \tag{VII.2}$$

then we must have $\tau = t_\ell$, i.e., $t \bmod \ell$ is obtained. The addition in the formula denotes point addition on the curve. The value of τ satisfying Equation (VII.2) is unique since ℓ is prime and $P \neq \mathcal{O}$.

To determine such a value of τ, assume for the time being that all values $\tau \in \{0, 1, \ldots, \ell - 1\}$ are tried in turn. First, the x-coordinates on both sides of Equation (VII.2) are computed. The x-coordinates of the point multiples $[q_\ell](x, y)$ and $[\tau](x^q, y^q)$, for the given prime ℓ and the value of τ being tested, are rational functions of x and y, involving the division polynomials (see Section III.4). The point addition formulae are used to symbolically compute the x-coordinate of $(x^{q^2}, y^{q^2}) + [q_\ell](x, y)$. By clearing denominators and, if necessary, eliminating powers of y higher than one by reducing modulo the curve equation (either $y^2 = x^3 + ax + b$ or $y^2 = xy + x^3 + a_6$), an equation of the form $a(x) - yb(x) = 0$ or $y = a(x)/b(x)$ results. This, in turn, can be substituted into the curve equation to eliminate y and give an equation of the form $h_X(x) = 0$. A crucial observation in determining the complexity of the procedure is that, since the postulated point P satisfying Equation (VII.2) is in $E[\ell]^*$, all polynomial computations can be carried out modulo the division polynomial f_ℓ, which is of degree $O(\ell^2)$. In particular, the polynomials

$x^{q^2}, y^{q^2}, x^q, y^q$ are reduced, using f_ℓ and the curve equation, from degree exponential in $\log q$ to degree polynomial in this parameter. The degree of $h_X(x)$ is therefore $O(\ell^2)$.

To check if $h_X(x) = 0$ has a solution for the x-coordinate of a point in $E[\ell]^*$, the greatest common divisor of h_X and f_ℓ is computed. If the GCD is one, then there is no solution in $E[\ell]^*$ which satisfies Equation (VII.2), and the next value of τ is tried. If the GCD is non-trivial, then there exists a point in $E[\ell]^*$ such that

$$(x^{q^2}, y^{q^2}) + [q_\ell](x, y) = \pm[\tau](x^q, y^q). \tag{VII.3}$$

The sign of the point on the right-hand side of the equation is ambiguous since the x-coordinates are the same for either sign. To determine the sign, assume it to be plus in Equation (VII.3). The y-coordinates of both sides of the equation are computed and, as with the x-coordinates, the denominators cleared and the y variable eliminated to give an equation of the form $h_Y(x) = 0$, with h_Y reduced to degree $O(\ell^2)$. Again, if $\gcd(h_Y, f_\ell) \neq 1$, there is a point satisfying the equation and the correct sign is plus; it is minus otherwise.

Notice that for a given τ, the procedure actually tests $\pm\tau$ and it is only necessary to have τ run through $0 \leq \tau \leq (\ell - 1)/2$ (the case $\tau = 0$ will require special treatment, which is discussed below). Generally the points of $E[\ell]$ have coordinates in an extension field of $\mathbb{F}_q$. Actual computation of these points, which would in general be very difficult, is avoided by the GCD computations.

To examine the complexity of the algorithm, we note that the bulk of the computation is taken up with finding $x^q, y^q, x^{q^2}, y^{q^2}$ (suitably reduced modulo the curve equation) modulo f_ℓ, a polynomial of degree $O(\ell^2)$. In the case of x^q and x^{q^2}, these are exponentiation operations in the ring $\mathbb{F}_q[x]/\langle f_\ell(x)\rangle$, requiring $O(\log q)$ multiplications in the ring. The modulus is of degree $O(\ell^2) = O(\log^2 q)$. Hence, assuming no fast multiplication routines are used, each such ring multiplication requires $O(\log^4 q)$ multiplications of elements of $\mathbb{F}_q$, each requiring in turn $O(\log^2 q)$ bit operations. The complexity of the y^q, y^{q^2} computations is similar, involving also reductions modulo the curve equation that do not affect the asymptotics. Notice that x^q, y^q, x^{q^2} and y^{q^2} are computed once for each prime ℓ and used for all the values of τ tried for that prime. Therefore, the number of bit operations needed for obtaining the trace modulo a single prime ℓ is $O(\log^7 q)$. Since the number of such primes is $O(\log q)$ (in fact, $O(\log q/\log\log q)$), the overall complexity for determining the group order is $O(\log^8 q)$ bit operations.

If fast polynomial and field arithmetic (as briefly mentioned in Chapter II) is used, then multiplication in $\mathbb{F}_q[x]/\langle f_\ell(x)\rangle$ takes $O(\log^{2+\epsilon} q)$ field multiplications, while field multiplications take $O(\log^{1+\epsilon} q)$ bit operations. The overall complexity, therefore, reduces to $O(\log^{5+\epsilon})$ bit operations. These gains, however, are mostly theoretical, and hard to realize in practical implementations.

In practice, $\log q$ is usually not sufficiently large to benefit from the fastest asymptotic improvements, yet it is large enough to make the naive implementation unacceptably slow. Intermediate solutions, as discussed in Chapter II, can be used (e.g., Karatsuba multiplication), but they will generally not suffice for the parameter ranges of practical interest. Better solutions will be sought, aimed mainly at finding a substitute for f_ℓ, of degree linear rather than quadratic in ℓ. These better solutions, in turn, could also benefit from fast arithmetic if the parameters of the problem so justified it.

The basic Schoof algorithm is summarized below.

ALGORITHM VII.1: **Basic Schoof Algorithm**

INPUT: An elliptic curve E over a finite field $\mathbb{F}_q$.
OUTPUT: The order of $E(\mathbb{F}_q)$.
1. $M \leftarrow 2$, $\ell \leftarrow 3$ and $S \leftarrow \{(t \pmod 2), 2)\}$.
2. While $M < 4\sqrt{q}$ do:
3. For $\tau = 0, \ldots, (\ell-1)/2$ do:
4. Using the formulae above check whether, for $P \in E[\ell]$, $\varphi^2(P) + [q]P = \pm[\tau]\varphi(P)$.
   Exactly one such τ will pass this test.
5. $S \leftarrow S \cup \{(\tau, \ell)\}$ or $S \leftarrow S \cup \{(-\tau, \ell)\}$, as appropriate.
6. $M \leftarrow M \times \ell$.
7. $\ell \leftarrow \mathrm{nextprime}(\ell)$.
8. Recover t using the set S and the CRT.
9. Return $q+1-t$.

In the above algorithm nextprime(ℓ) is a function which returns the smallest prime number larger than ℓ.

The computations discussed above are considered now in slightly greater detail to illustrate the technique further. For the discussion, we focus on the case of characteristic two, as the formulae for division polynomials $f_m(x)$ and point multiples based on them can be used without having to consider various parity cases. We show the computations for the x-coordinate only, the ones for y being similar.

Assume first that for no point $P \in E[\ell]^*$ is it true that $\varphi^2(P) = \pm[q_\ell]P$. We search for a $\tau \in \mathbb{F}_\ell$ such that[1]

$$\varphi^2(P) + [q_\ell]P = \pm[\tau]\varphi(P),\ 1 \leq \tau \leq (\ell-1)/2,\ q = 2^n.$$

The x-coordinates of the two sides of the equation can be computed using the point addition formulae from Section III.3.2, and the point multiplication

[1] Here, and later, we will slightly abuse notation and write $[\tau]Q$ for $\tau \in \mathbb{F}_\ell$ and a point Q. Since the multiplication-by-τ map was formally defined for $\tau \in \mathbb{Z}$, what is meant is $[\tau']Q$ for any integer τ' in the congruence class modulo ℓ naturally associated with τ. This will always be applied to points $Q \in E[\ell]$, so there is no ambiguity.

formulae from Section III.4 for the characteristic two case. The x-coordinates are, respectively,

$$(\pm[\tau]\varphi(P))_X = x^q + \frac{f_{\tau-1}^q f_{\tau+1}^q}{f_\tau^{2q}},$$

and

$$\left(\varphi^2(P) + [q_\ell]P\right)_X = x^{q^2} + x + \frac{f_{q_\ell-1}f_{q_\ell+1}}{f_{q_\ell}^2} + \lambda^2 + \lambda,$$

where

$$\lambda = \frac{(y^{q^2} + y + x)xf_{q_\ell}^3 + f_{q_\ell-2}f_{q_\ell+1}^2 + (x^2 + x + y)(f_{q_\ell-1}f_{q_\ell}f_{q_\ell+1})}{xf_{q_\ell}^3(x + x^{q^2}) + xf_{q_\ell-1}f_{q_\ell}f_{q_\ell+1}},$$

the subscript X on the bracket indicates the x-coordinate, and x is the assumed argument of the various division polynomials f_m. Notice that since the latter have coefficients in $\mathbb{F}_q$, we have $f_m(x^q) = f_m(x)^q$. The case of $q_\ell = 1$ might be handled separately, since the above expression for λ involves $f_{q_\ell-2}$, which has not been defined for $q_\ell = 1$. The case involves the addition of the points (x^{q^2}, y^{q^2}) and (x, y) which is straightforward.

The powers of y in the equation are reduced modulo the curve equation $y^2 + xy + x^3 + a_6 = 0$, to yield polynomials of degree at most one in y. Both sides of the expressions are multiplied by the LCM of the polynomials in the denominators to give a relationship of the form $a(x) + yb(x) = 0$, where in forming $a(x)$ and $b(x)$, polynomial computations are carried out modulo $f_\ell(x)$. In fact, the reductions modulo f_ℓ and the curve equation must occur in an interleaved fashion, as we do not wish to manipulate polynomials of degree exponential in $\log q$. The relation $y = a(x)/b(x)$ is substituted into the curve equation to give

$$h_X(x) = a^2(x) + xa(x)b(x) + b^2(x)x^3 + b^2(x)a_6 = 0.$$

If $\gcd(h_X, f_\ell) \neq 1$ then a point $P \in E^*[\ell]$ exists whose x-coordinate satisfies Equation (VII.1). In this case the y-coordinates of the two sides are checked in a similar manner to determine the correct sign.

We comment briefly on the case where there is a point $P \in E^*[\ell]$ such that $\varphi^2(P) = \pm[q_\ell]P$. This case, which was excluded in the discussion above, arises if

$$\gcd\left((x^{q^2} + x)f_{q_\ell}^2 + f_{q_\ell-1}f_{q_\ell+1}, f_\ell\right) \neq 1.$$

Clearly $t \equiv 0 \pmod{\ell}$ if and only if $\varphi^2(P) = -[q_\ell]P$. This condition can be tested by checking the y-coordinate, as before. If the condition is not verified, then $\varphi^2(P) = +[q_\ell]P$ for some point P, and, by the characteristic Equation (VII.2), we have

$$[2q_\ell]P - [\tau]\varphi(P) = \mathcal{O},$$

or

$$\varphi(P) = [\frac{2q_\ell}{\tau}]P.$$

Applying the Frobenius map to both sides of the equation, and again the equality satisfied by P, it follows that $\tau^2 \equiv 4q \pmod{\ell}$ and, thus, that q has a square root modulo ℓ, say $w \in \mathbb{F}_\ell$. Thus $\varphi(P) = \pm[w]P$ and the cases are distinguished as before. Assume $\varphi(P) = [w_0]P$, with $w_0 \in \{+w, -w\}$. Then, we set $t_\ell \equiv 2w_0 \pmod{\ell}$. Notice that in this case, the Frobenius map is said to have an eigenvalue $w_0 \in \mathbb{F}_\ell$, which we encountered while handling a rather special case. The existence of such eigenvalues of the Frobenius map, in more general cases, forms the basis of the Elkies improvement of the basic Schoof algorithm. Such eigenvalues will exist when $t^2 - 4q$ is a square in $\mathbb{F}_\ell$ and they lead to the existence of a factor of degree $(\ell - 1)/2$ of the division polynomial f_ℓ. Reducing the equations of this section modulo this polynomial, rather than the division polynomial, will lead to very significant computational savings.

VII.2. Beyond Schoof

To improve the computational efficiency of the basic Schoof algorithm, several useful techniques have evolved, owing in large part to Atkin and Elkies. Many of the crucial observations of these two foremost researchers have only recently appeared in non-electronic form.[2] Descriptions of their important contributions are now available in [**142**] and [**40**], and in two key doctoral theses [**110**] and [**81**].

The evolution of these techniques has proved remarkably successful, culminating in the determination of the order of the group of rational points on a randomly chosen curve, in the case of a prime field for a prime of some five hundred decimal digits ([**108**], [**83**]) and in the case of a field of characteristic two for $\mathbb{F}_{2^{1301}}$ ([**81**], [**84**]) and $\mathbb{F}_{2^{1663}}$ (as reported in a recent electronic communication by Lercier).

The techniques of Atkin and Elkies depend on whether the roots of the characteristic equation of the Frobenius map,

$$\mathcal{F}_\ell(u) = u^2 - t_\ell u + q_\ell = 0,$$

taken modulo a prime ℓ, lie in $\mathbb{F}_\ell$ or not, i.e., whether the discriminant $\Delta_t = t^2 - 4q$ is a square modulo ℓ or not. In the case that it is, the prime ℓ is said to be an Elkies prime and in the case that it is not, an Atkin prime. Of course, since the trace t is unknown – it is what we are trying to determine – one has to resort to other techniques to determine which case is in effect for a given prime ℓ. It will turn out that the modular polynomials may be used for this determination. While these polynomials are defined over the complex numbers, $\mathbb{C}$, as noted in Chapter III, they have integer coefficients and hence can be interpreted over any field. As will be noted in Proposition VII.2 below, the splitting type of the ℓth modular polynomial $\Phi_\ell(x, y)$

[2] As Birch [**11**] comments, Atkin's way is 'to make his work known by bush telegraph, via e-mail, or as quoted by others.'

over the ground field $\mathbb{F}_q$, with the j-invariant of the curve substituted for one of the variables, determines whether Δ_t is a square in $\mathbb{F}_\ell$ or not, and thus whether ℓ is an Elkies or Atkin prime. While the computation of the modular polynomials is a challenging task, this is a convenient test to decide, for a given curve and prime ℓ, which of the two possibilities is in effect. In the following subsections, we outline the Elkies and Atkin approaches, and how they determine information on t modulo ℓ. We will provide more detail later on, after covering some additional mathematical background.

VII.2.1. Elkies primes. We assume ℓ is an odd prime and ℓ is not p, the characteristic of the field. When ℓ is an Elkies prime, the discriminant Δ_t is a square in $\mathbb{F}_\ell$, and the characteristic polynomial $\mathcal{F}_\ell$ of φ has two roots, say λ and μ, in $\mathbb{F}_\ell$, which are eigenvalues of the Frobenius map modulo ℓ. The determination of whether Δ_t is a square in $\mathbb{F}_\ell$ or not can be deduced from the splitting type of the ℓth modular polynomial, in a manner to be considered in the next section. Assume for convenience that $\lambda \neq \mu$. The case of $\lambda = \mu \in \mathbb{F}_\ell$ was discussed briefly at the end of Section VII.1. It corresponds to $t \equiv \pm 2\sqrt{q} \pmod{\ell}$, and we will further elaborate on it later. The set of ℓ-torsion points, $E[\ell]$, has two of its $\ell + 1$ cyclic subgroups, say C_1 and C_2, that are stable under the Frobenius endomorphism, i.e., $\varphi(P_1) = \lambda P_1$ for all $P_1 \in C_1$ and $\varphi(P_2) = \mu P_2$ for all $P_2 \in C_2$. The characteristic polynomial factors over $\mathbb{F}_q$ as

$$\mathcal{F}_\ell(u) = u^2 - tu + q = (u - \lambda)(u - \mu)\,.$$

The aim in this case is to determine one of the roots, say λ, since then

$$t \equiv \lambda + \frac{q}{\lambda} \pmod{\ell}. \qquad \text{(VII.4)}$$

To find such an eigenvalue we could test for a point $P = (x, y)$ and a value $\lambda \in \{1, 2, \ldots, \ell - 1\}$, such that

$$(x^q, y^q) = [\lambda](x, y).$$

Notice that in this case, computation of x^{q^2} and y^{q^2} is not required. However the computation of x^q is of the same asymptotic complexity, and still an onerous task (computing $x^{q^2} \pmod{f_\ell}$ is only about twice as complex as computing $x^q \pmod{f_\ell}$). In the case of an Elkies prime, it will be shown how information derived from the modular polynomials will lead to the determination of a factor of degree $(\ell - 1)/2$ of the division polynomial f_ℓ, which can be used for the reductions needed in testing a potential eigenvalue λ. An outline of the method is given here with the details to follow in succeeding sections.

To construct such a polynomial, a curve isogenous to E, say E_1, is sought such that the isogeny is of degree ℓ, i.e., the kernel of the isogeny is of cardinality ℓ. The kernel of the isogeny, say C (one of the subgroups C_1 or

C_2 above), is stable under the action of the Frobenius map and hence the polynomial

$$F_\ell(x) = \prod_{\pm P_i \in C\setminus\{\mathcal{O}\}} (x - (P_i)_X) \tag{VII.5}$$

is defined over the field of definition of the curve, where the product includes only one of each pair $\pm P_i$, since both points have the same x-coordinate. The degree of $F_\ell(x)$ is $(\ell - 1)/2$.

If the original curve has j-invariant j, then the isogenous curves have j-invariants that are zeros of the ℓth modular polynomial $\Phi_\ell(x, j)$. For the case of an Elkies prime, one of the two j-invariants (zeros of the modular polynomial) that are in the ground field $\mathbb{F}_\ell$ is chosen. Determining such an isogenous curve and the polynomial $F_\ell(x)$ will be the major task of much of the remainder of the chapter.

Once $F_\ell(x)$ is determined, it can be used to efficiently compute for which λ it is true that

$$(x^q, y^q) = [\lambda](x, y), \ (x, y) \in C\,.$$

This procedure is similar to the main computation in the original Schoof algorithm, but with polynomial operations modulo $F_\ell(x)$ instead of $f_\ell(x)$. Given λ, the value of $t \pmod{\ell}$ is then uniquely determined by Equation (VII.4).

VII.2.2. Atkin primes. As noted, ℓ is an Atkin prime when $t^2 - 4q$ is not a square modulo the prime ℓ, which, as for the case of Elkies primes, will be determined from information on the splitting type of the modular polynomial $\Phi_\ell(x, j)$. It will be noted in the next section that this information also determines a subset of possible values of the trace modulo ℓ. This subset is of size $\phi_{\mathrm{Eul}}(r)$, where ϕ_{Eul} denotes the Euler totient function and $r \le \ell + 1$ is an integer to be discussed later. This subset is contrasted to the exact value of $t \pmod{\ell}$ obtained in the case of an Elkies prime. A BSGS/CRT procedure is then applied to merge the information found from both types of primes to determine the exact value of t, as discussed in Section VII.9.

VII.2.3. Outline of the SEA algorithm. The improvements of Elkies and Atkin to the basic Schoof algorithm are generally referred to as the Schoof–Elkies–Atkin (SEA) algorithm. The algorithm is outlined here, and treated in greater detail in the following sections.

ALGORITHM VII.2: **Schoof–Elkies–Atkin (SEA) Algorithm**

INPUT: An elliptic curve E over a finite field $\mathbb{F}_q$.
OUTPUT: The order of $E(\mathbb{F}_q)$.
1. $M \leftarrow 1$, $\ell \leftarrow 2$, $A \leftarrow \{\}$ and $E \leftarrow \{\}$.
2. While $M < 4\sqrt{q}$ do:
3. Decide whether ℓ is an Atkin or Elkies prime, by finding the splitting type of the modular polynomial.

```
4.        If ℓ is an Elkies prime then do:
5.            Determine the polynomial F_ℓ(x) above.
6.            Find an eigenvalue, λ, modulo ℓ.
7.            t ← λ + q/λ (mod ℓ).
8.            E ← E ∪ {(t, ℓ)}.
9.        Else do:
10.           Determine a (small) set T such that t (mod ℓ) ∈ T.
11.           A ← A ∪ {(T, ℓ)}.
12.       M ← M × ℓ.
13.       ℓ ← nextprime(ℓ).
14. Recover t using the sets A and E, the CRT and BSGS.
15. Return q + 1 − t.
```

Although we are yet to discuss how some of the crucial steps of this algorithm are actually performed, we can comment on the asymptotic computational advantage of some of the outlined improvements. Most notable are the gains in the processing of Elkies primes. Here, the bottleneck is the computation of x^q and y^q modulo the polynomial $F_\ell(x)$ and the curve equation. With an analysis similar to that of the Schoof algorithm, we can bound the computational complexity of such computations at $O(\ell^5)$ bit operations if naive arithmetic is implemented, or $O(\ell^{3+\epsilon})$ using fast arithmetic (compare with $O(\ell^7)$ and $O(\ell^{4+\epsilon})$, respectively, for the Schoof algorithm). To take advantage of this speed up, the complexity of *obtaining* $F_\ell(x)$ must not exceed the complexity of *using* it. Thus, the complexity of constructing $F_\ell(x)$ should not exceed $O(\ell^3)$ operations in $\mathbb{F}_q$ when naive arithmetic is used, or $O(\ell^{2+\epsilon})$ in the case of fast arithmetic. In particular, the straightforward approach of attempting to factor $f_\ell(x)$ to obtain $F_\ell(x)$ does not seem to work, as one of the first steps in such a factorization would involve a computation of the type $x^q \pmod{f_\ell(x)}$, precisely what we are trying to avoid.

Since the number of Elkies primes processed will be $O(\log q)$, the overall complexity of the Elkies portion of the algorithm is $O(\log^6 q)$ for naive arithmetic, or $O(\log^{4+\epsilon})$ for fast arithmetic.

Turning to the Atkin portion of the algorithm, it will transpire that this portion is actually of exponential asymptotic complexity. As will be discussed in Section VII.8, the complexity-theory way out of this unpleasant predicament is to deal only with Elkies primes, processing enough of them to build up a modulus M from which the trace of Frobenius can be determined. Thus, the asymptotic analysis for the Elkies portion above would apply to the whole algorithm. In practice, however, this approach has several disadvantages, and it will turn out that a subset of the Atkin primes can be used to advantage for the ranges of field sizes of interest in cryptographic applications. The subset is carefully chosen to balance the overall complexity, meaning that some Atkin primes (eventually a majority, as q increases) will not be processed in Steps 10–11, and their contribution not counted in Step 12.

In the next sections, we provide more details on the main steps of the SEA algorithm. In particular, we elaborate on steps 3, 5, 6, 10 and 14. In Section VII.3 modular polynomials and their splitting types are discussed, providing a method to implement Step 3. Techniques for determining the polynomials F_ℓ, which will occupy most of the effort, will be treated in Sections VII.4 (odd characteristic) and VII.5 (characteristic two), respectively. The discussion on odd characteristic in Section VII.4 will focus on fields of large prime order p, a departure from the more general case $q = p^n$, but one that is most relevant in practice. After describing the computation of the polynomial F_ℓ, we return to the Elkies procedure (how the polynomial is used in Step 6) in Section VII.7, and to the details of the Atkin procedure (Step 10) in Section VII.8. In Section VII.9 we show how the information from Elkies primes and Atkin primes is combined (Step 14). Finally, in Section VII.10 some examples are presented, and in Section VII.11 other recent algorithms are briefly discussed.

VII.3. More on the Modular Polynomials

The modular polynomials $\Phi_\ell(x, y)$, for elliptic curves defined over $\mathbb{C}$, were introduced in Section III.8. The properties of these polynomials play a central role in the point counting algorithms. The polynomials are symmetric and have integer coefficients, thus can be interpreted over any field. In this and succeeding sections, we will need to consider the polynomial $\Phi_\ell(x, j)$ and its factorizations over a ground field $\mathbb{F}_q$, where $j \in \mathbb{F}_q$ is the j-invariant of a given curve.

The following proposition is of interest:

PROPOSITION VII.1 (see [**142**]). *Let E be a non-supersingular elliptic curve over $\mathbb{F}_q$, with j-invariant $j \neq 0, 1728$. Then*

(i) *the polynomial $\Phi_\ell(x, j)$ has a zero $\tilde{\jmath} \in \mathbb{F}_{q^r}$ if and only if the kernel C of the corresponding isogeny*

$$\phi : E \longrightarrow E/C$$

is a one dimensional eigenspace of φ^r in $E[\ell]$ where φ is the Frobenius endomorphism of E (here, $j(E/C) = \tilde{\jmath}$),

(ii) *the polynomial $\Phi_\ell(x, j)$ splits completely in $\mathbb{F}_{q^r}[x]$ if and only if φ^r acts as a scalar matrix on $E[\ell]$.*

The following proposition, from Schoof [**142**] and Lercier [**81**], is critical to both the Atkin and Elkies improvements of the basic Schoof algorithm. The proposition is attributed by both [**142**] and [**81**] to unpublished electronic communications by Atkin. Recall from Section III.8 that $\Phi_\ell(x, j)$ is of degree $\ell + 1$ in x.

PROPOSITION VII.2 ([**142**] [**81**]). *Let E be a non-supersingular elliptic curve over $\mathbb{F}_q$ with j-invariant $j \neq 0, 1728$. Let $\Phi_\ell(x, j) = h_1 h_2 \cdots h_s$ be the factorization of $\Phi_\ell(x, j) \in \mathbb{F}_q[x]$ as a product of irreducible polynomials. Then there are the following possibilities for the degrees of $h_1, h_2, \ldots, h_s$:*

(i) *either 1 and ℓ (in which case we set $r = \ell$) or $1, 1, \ldots, 1$ (in which case we set $r = 1$) – in either situation ℓ divides the discriminant, i.e., $t^2 - 4q \equiv 0 \pmod{\ell}$;*

(ii) *$1, 1, r, r, \ldots, r$ – in this case $t^2 - 4q$ is a square mod ℓ, r divides $\ell - 1$ and φ acts on $E[\ell]$ as a matrix*

$$\begin{pmatrix} \lambda & 0 \\ 0 & \mu \end{pmatrix}$$

with $\lambda, \mu \in \mathbb{F}_\ell^*$;

(iii) *$r, r, \ldots, r$ for some $r > 1$ – in this case $t^2 - 4q$ is not a square modulo ℓ, r divides $\ell + 1$ and φ acts on $E[\ell]$ as a 2×2 matrix whose characteristic polynomial is $\mathcal{F}_\ell$, which is irreducible over $\mathbb{F}_\ell$.*

In all three cases r is the order of φ in $PGL_2(\mathbb{F}_\ell)$ and the trace t of φ satisfies the equation

$$t^2 = q(\zeta + 2 + \zeta^{-1}) \tag{VII.6}$$

over $\mathbb{F}_\ell$, for some primitive rth root of unity $\zeta \in \overline{\mathbb{F}}_\ell$.

First, we note that the proposition provides a way to classify ℓ as an Elkies or Atkin prime, through the factorization of $\Phi_\ell(x, j)$. Cases (i) and (ii) of this proposition correspond to the prime ℓ being an Elkies prime, with Case (i) corresponding to the case of $\mathcal{F}_\ell$ having a double root. Case (iii) of the proposition corresponds to an Atkin prime.

The proof of Equation (VII.6) is straightforward. Since r is the order of φ in $PGL_2(\mathbb{F}_\ell)$, r is the smallest integer such that $\lambda^r = \mu^r$, i.e. the smallest integer such that φ^r is represented by a scalar multiple of the identity matrix. Since $\lambda\mu = q$, we have $\lambda^{2r} = q^r$ and hence $\lambda^2 = \zeta q$ for a primitive rth root of unity in some extension field of $\mathbb{F}_\ell$. Thus $t^2 = (\lambda + q/\lambda)^2 = q(\zeta + 2 + \zeta^{-1})$. In Case (i) of the proposition, take $\zeta = 1$.

Equation (VII.6) is of particular importance for the Atkin algorithm, since it will limit the number of possible values that the trace can have, for the given prime. This number is $\phi_{\mathrm{Eul}}(r)$, the number of primitive rth roots of unity. Since the proposition states that r divides $\ell + 1$ in this case, all these roots are in $\mathbb{F}_{\ell^2}$. Each pair ζ, ζ^{-1} determines one value of t^2, or two values of t when $t_\ell = t \pmod{\ell} \neq 0$, for a total of $\phi_{\mathrm{Eul}}(r)$ possible values of t_ℓ. For example, we get $t_\ell = 0$ when $\zeta = -1$, which can occur only when $r = 2$. In this case, $\phi_{\mathrm{Eul}}(r) = 1$, consistent with the single possible value $t_\ell = 0$. In another example, when $r = 4$ we must have $\zeta + \zeta^{-1} = 0$. Then, $t^2 \equiv 2q \pmod{\ell}$, allowing only two possible values for t_ℓ.

Notice that Proposition VII.2 gives both Steps 3 and 10 of Algorithm VII.2. Notice also that Step 3 does not require the factorization of the modular polynomial. It is sufficient to note that the degree of

$$\gcd\left(x^q - x, \Phi_\ell(x, j)\right)$$

is 0, 1, 2 or $\ell + 1$, where degrees 1, 2 and $\ell + 1$ correspond to an Elkies prime and degree 0 to an Atkin prime.

It is further shown by Schoof [**142**] that the parity of the number of irreducible factors, s, of Φ_ℓ when q is a large prime p is easily obtained from the Legendre symbol:

$$(-1)^s = \left(\frac{p}{\ell}\right).$$

The modular polynomial $\Phi_\ell(x, y)$ over $\mathbb{C}$ has integer coefficients, as noted in Section III.8. We recall that, over $\mathbb{C}$, the roots of $\Phi_\ell(x, j(\tau))$ are given by

$$j\left(\frac{\tau + b}{\ell}\right) \quad \text{for } 0 \le b < \ell, \quad \text{and} \quad j(\ell\tau). \tag{VII.7}$$

These roots correspond to the $\ell + 1$ matrices in the set

$$S_\ell^* = \left\{ \begin{pmatrix} a & b \\ 0 & d \end{pmatrix} : a, b, d \in \mathbb{Z}_{\ge 0}, \;\; ad = \ell, \;\; 0 \le b < d \right\}.$$

If $j(\tau)$ is the j-invariant of an elliptic curve E over $\mathbb{C}$, then each such root of $\Phi_\ell(x, j(\tau))$ is the j-invariant of a curve isogenous to E under an isogeny of degree ℓ.

It turns out that a similar result holds for curves over finite fields. In this case, the modular polynomials are interpreted by reducing their coefficients modulo the characteristic of the field. The following theorem gives the one-to-one correspondence between roots of the modular polynomials, possible j-invariants of isogenous curves, and subgroups of the ℓ-torsion points, over a finite field $\mathbb{F}_q$.

THEOREM VII.3 (see [**74**], [**110**]). *Let E be an elliptic curve over $\mathbb{F}_q$, ℓ a prime different from the field characteristic, and C_i, $1 \le i \le \ell + 1$, the subgroups of exact order ℓ of $E(\overline{\mathbb{F}}_q)$. Let $\Phi_\ell(x, y)$ be the ℓth modular polynomial, taken over $\mathbb{F}_q$. Then, all the roots of $\Phi_\ell(x, j(E))$ are given by the j-invariants of E/C_i, $1 \le i \le \ell + 1$.*

The difficulty of working with these modular polynomials, whose coefficients can become extremely large, has been noted in Section III.8. Although the polynomials we require have coefficients reduced modulo the characteristic of $\mathbb{F}_q$, they are often computed, initially, over $\mathbb{Z}$. In several places in the literature, more manageable alternatives to these polynomials, with the essential splitting properties preserved, have been suggested (see, for instance, [**108**], [**40**] and [**110**]). However, even the modified modular polynomials require great care to compute. One set of alternatives which we have looked at in

Section III.8 is the polynomials $G_\ell(x, y)$, due to Müller. These polynomials have the same splitting type over $\mathbb{F}_q$ as $\Phi_\ell(x, y)$ (see Theorem III.17), and can be used in lieu of the modular polynomials to determine whether Δ_t is a square in $\mathbb{F}_\ell$ or not.

The splitting type of a modular polynomial and the information in Proposition VII.2 are related to the factorization of the characteristic polynomial of the Frobenius map in $\mathbb{F}_\ell$. When it is determined that ℓ is an Elkies prime, this information will also bear on the factorization of the division polynomials f_ℓ, as will be shown below and in the next two sections. These factorizations, in turn, are related to the action of powers of the Frobenius endomorphism on subgroups of the ℓ-torsion points of E. A few comments on these matters are given here. They duplicate some of the material already given, but the interpretation of the results in the different setting is a worthwhile diversion. Much of the discussion is from [**81**].

Suppose the field of definition of the curve has characteristic p. Recall from Lemma III.8 that if $\gcd(\ell, p) = 1$ then the structure of the group of ℓ-torsion points is

$$E[\ell] \cong (\mathbb{Z}/\ell\mathbb{Z}) \times (\mathbb{Z}/\ell\mathbb{Z}).$$

The group $E[\ell]$ is generated by two points, say P_1 and P_2, and $E[\ell]$ contains the $\ell + 1$ subgroups $C_1 = \langle P_1 \rangle$, $C_2 = \langle P_2 \rangle$ and $C_i = \langle P_1 + (i-2)P_i \rangle$ for $i = 3, 4, \dots, \ell + 1$. The subgroups share the point at infinity $\mathcal{O}$ and their union is $E[\ell]$. Each such subgroup is the kernel of an isogeny of degree ℓ, and the j-invariants of the isogenous curves are given by the roots of Φ_ℓ as discussed previously, Theorem VII.3.

Consider the action of the Frobenius map, φ, on the subgroups of $E[\ell]$. As before let λ and μ be the roots of the characteristic polynomial $\mathcal{F}_\ell(u) = u^2 - t_\ell u + q_\ell$ in $\mathbb{F}_\ell$. Let e_1 and e_2 be the orders of λ and μ, respectively. Three cases are distinguished from the above discussion, which we elaborate on. The cases correspond to those of Proposition VII.2:

(i) $\lambda = \mu \in \mathbb{F}_\ell$ (Case (i) of Proposition VII.2), i.e., the characteristic polynomial is $\mathcal{F}_\ell(u) = (u - \lambda)^2$. In this case $t = \pm 2\sqrt{q} \pmod{\ell}$ and there exist a point P_1 and subgroup C_1 such that $\varphi(P_1) = [\lambda]P_1$, $\varphi(C_1) = C_1$. Also, there exists a point P_2 not in C_1 such that $\varphi(P_2) = [\lambda]P_2 + [k]P_1$, for some $k \in \mathbb{F}_\ell$. There are two subcases:
   (a) if $k \neq 0$ then $\varphi^{e_1}(P_2) = P_2$ and
   $$C_i \subset E(\mathbb{F}_{q^{e_1}}) \text{ and } \varphi^\ell(C_i) = C_i$$
   for $i = 2, 3, \dots, \ell + 1$ – this corresponds to the splitting type $1, \ell$;
   (b) if $k = 0$ then $C_i \subset E(\mathbb{F}_{q^{e_1}})$ and $\varphi(C_i) = C_i$, $i = 2, 3, \dots, \ell + 1$ – this corresponds to the splitting type $1, 1, \cdots, 1$.

(ii) $\lambda \neq \mu$, $\lambda, \mu \in \mathbb{F}_\ell$ (Case (ii) of Proposition VII.2). In this case $\mathcal{F}_\ell(u) = (u - \lambda)(u - \mu)$ and there exists points P_1, $P_2 \in E[\ell]$ such that
$$\varphi(P_1) = [\lambda]P_1, \quad \varphi(P_2) = [\mu]P_2.$$

It is clear that P_1 and P_2 must lie in different subgroups of $E[\ell]$, and we can take $C_1 = \langle P_1 \rangle$, $C_2 = \langle P_2 \rangle$, and $E[\ell] = C_1 \times C_2$. From the assumed orders of the eigenvalues it follows that

$$\varphi^{e_1}(P_1) = P_1, \quad \varphi^{e_2}(P_2) = P_2$$

and the coordinates of the points in C_1 lie in $\mathbb{F}_{q^{e_1}}$ and those of C_2 lie in $\mathbb{F}_{q^{e_2}}$ (where e_1 and e_2 divide $\ell - 1$). Any point $Q \in E[l]$ can be expressed as $[m_1]P_1 + [m_2]P_2$, for some $m_1, m_2 \in \mathbb{F}_\ell$. If $e = \text{lcm}(e_1, e_2)$, then the coordinates of all points of $E[\ell]$ lie in $\mathbb{F}_{q^e}$. Thus, for any integer s

$$\varphi^s(Q) = [\lambda^s m_1]P_1 + [\mu^s m_2]P_2$$

and the smallest integer k for which $\varphi^k(Q)$ is a multiple of Q is the order of λ/μ in $\mathbb{F}_\ell$, say r. It follows that

$$\varphi^r(C_i) = C_i, \ i = 3, 4, \ldots, \ell + 1,$$

while for $i = 1$ and 2 we have $\varphi(C_1) = C_1$ and $\varphi(C_2) = C_2$. This is the case where ℓ is an Elkies prime and the value of the trace modulo ℓ is resolved by the computational method discussed later in the chapter.

(iii) $\lambda \neq \mu$, $\lambda, \mu \in \mathbb{F}_{\ell^2} - \mathbb{F}_\ell$ (Case (iii) of Proposition VII.2). This case corresponds to ℓ being an Atkin prime and $\mathcal{F}_\ell(u)$ being irreducible over $\mathbb{F}_\ell$. We can write $\lambda = \mu^\ell$. As with the previous case, if $e = \text{lcm}(e_1, e_2)$ then $C_i \subset E(\mathbb{F}_{q^e})$. Also, if r is the order of λ/μ in $\mathbb{F}_{\ell^2}$ then for the subgroups C_i of $E[\ell]$ one has

$$\varphi^r(C_i) = C_i \ , \ i = 1, 2, \ldots, \ell + 1.$$

In this case, since there are exactly $\phi_{\text{Eul}}(r)$ elements in $\mathbb{F}_{\ell^2}$ of order exactly r, there are exactly $\phi_{\text{Eul}}(r)$ possibilities for the value of the trace, the ambiguity being resolved later in the computation by the merge-and-sort or BSGS routine noted.

It is finally noted that, from the above discussion, it follows that for a non-supersingular elliptic curve E over $\mathbb{F}_q$, the powers φ^{e_i} of the Frobenius map that leave the various subgroups of $E[\ell]$ stable are given by

$$e_i = \min\{\, k \in \mathbb{Z} : j(E/C_i) \in \mathbb{F}_{q^k} \}\,.$$

VII.4. Finding Factors of Division Polynomials through Isogenies: Odd Characteristic

Our purpose in this section and the next is to outline the development of the polynomial $F_\ell(x)$, a divisor of the ℓth division polynomial, $f_\ell(x)$, of degree $d = (\ell - 1)/2$, for the cases of odd and even characteristic, respectively. The variable d is reserved for $(\ell - 1)/2$ in what follows. In both cases the technique will be to determine a curve isogenous to the given curve, and obtain sufficient information about points in the kernel of the isogeny such that all the coefficients of the polynomial may be determined. The x-coordinates of points in the kernel will be the roots of $F_\ell(x)$.

It will be assumed throughout this section that the ground field is $\mathbb{F}_p$, with p a large prime. It is further assumed that ℓ is an Elkies prime, the only case for which such factors of the division polynomials are used in Algorithm VII.2. To determine the factor $F_\ell(x)$, of degree d, the steps will be as follows:

(i) Given a curve over $\mathbb{F}_p$ with j-invariant j, determine a j-invariant $\tilde{\jmath}$ of an isogenous curve, by determining a root of the modular polynomial $\Phi_\ell(x, j)$, i.e. find $\tilde{\jmath}$ such that $\Phi_\ell(\tilde{\jmath}, j) = 0$

(ii) For the given j-invariant $\tilde{\jmath}$, determine the coefficients $\tilde{a}$, $\tilde{b}$ of an isogenous curve,
$$Y^2 = X^3 + \tilde{a}X + \tilde{b},$$
with j-invariant $\tilde{\jmath}$.

(iii) From knowledge of the isogenous curves, and the kernel of the isogeny, compute the sum of the x-coordinates of the points in the kernel of the isogeny. From this last quantity and the two curves, derive the desired polynomial $F_\ell(x)$.

All the advanced techniques currently available in the literature for counting points on an elliptic curve over $\mathbb{F}_p$ perform these steps or simple variants of them. The rest of this section gives a glimpse into what is involved in these steps, following closely the treatment in [**142**]. In the latter reference, the approach is again attributed to unpublished electronic correspondence by Atkin.

Consider the curve
$$Y^2 = X^3 + aX + b$$
over $\mathbb{F}_p$. The j-invariant of this curve is
$$j(E) = 1728 \frac{4a^3}{4a^3 + 27b^2}. \tag{VII.8}$$
In the case of the Elkies prime ℓ, by Proposition VII.2, we have
$$\deg\left(\gcd(x^p - x, \Phi_\ell(x, j))\right) > 0,$$
and the roots of the GCD are in $\mathbb{F}_p$. Typically there will be two such roots, and we will designate one of them $\tilde{\jmath}$. Thus there will be a curve isogenous to E, E/C, with j-invariant $\tilde{\jmath} = j(E/C)$ such that the isogeny between E and E/C has kernel C of order ℓ. The Weierstrass equation of E/C will be of the form
$$Y^2 = X^3 + \tilde{a}X + \tilde{b}.$$
The next task is to determine the coefficients $\tilde{a}$ and $\tilde{b}$, from knowledge of a, b, $j(E)$, and $\tilde{\jmath}$.

For this development, a detour through the complex model of the elliptic curve will be required. The theory behind this detour is deep, and a full account of it is beyond the scope of the book. The basic facts that make it work are the following. Associated with the original curve E (its coefficients interpreted as integers) are complex parameters τ and $q = \exp(2\pi i\tau)$. From them, the various invariants $j(q)$, $\Delta(q)$, $E_4(q)$, $E_6(q)$, defined in Chapter III, are derived.[3] It can be established that these quantities reside in the ring of integers $\mathcal{O}_K$ of some number field K, and that there is a prime ideal $\mathfrak{B}$ of $\mathcal{O}_K$ with residue field $\mathbb{F}_p$, such that the reduction of these quantities modulo the ideal $\mathfrak{B}$ yields integers modulo p. In the case of the j-invariant $j(q)$, for instance, its residue modulo $\mathfrak{B}$ yields the j-invariant $j(E) \in \mathbb{F}_p$ of Equation (VII.8). Similarly, the complex quantities $E_4(q)$ and $E_6(q)$ have counterparts in $\mathbb{F}_p$, which will be denoted by $\overline{E}_4(q)$, $\overline{E}_6(q)$, respectively, to emphasize the distinction. Various relations will be established between the complex versions of these quantities, which carry over to $\mathbb{F}_p$. Ultimately, all the computations are actually done over $\mathbb{F}_p$, the complex model used only to justify some of the steps.

We start by recalling from Section III.7 the expressions for the following Eisenstein formal series in $\mathbb{Z}[[q]]$:

$$E_2(q) = 1 - 24\sum_{n=1}^{\infty}\frac{nq^n}{1-q^n}, \quad E_4(q) = 1 + 240\sum_{n=1}^{\infty}\frac{n^3q^n}{1-q^n},$$

$$E_6(q) = 1 - 504\sum_{n=1}^{\infty}\frac{n^5q^n}{1-q^n}.$$

In addition, the discriminant is expressed as

$$\Delta(q) = q\prod_{n=1}^{\infty}(1-q^n)^{24}.$$

We also recall the relationships

$$j(q) = \frac{E_4(q)^3}{\Delta(q)} = \frac{1}{q} + 744 + 196884q + 21493760q^2 + \cdots$$

and

$$\Delta(q) = \frac{E_4(q)^3 - E_6(q)^2}{1728}.$$

Denoting q times the derivative of a formal power series, $f(q) = \sum_n a_n q^n$, by

$$f'(q) = q\frac{df}{dq} = \sum\nolimits_n n a_n q^n$$

[3]In this section, we write $\Delta(\cdot)$, $j(\cdot)$, $E_4(\cdot)$, and $E_6(\cdot)$ as functions of q, rather than functions of τ, as done when these functions were defined in Chapter III. This is convenient, since various formal derivatives, taken with respect to q, will be required. While this notation is formally imperfect, the relation $q = \exp(2\pi i\tau)$ makes the functional relations unambiguous, and it helps reduce notation clutter. We will switch quite freely between the two notations.

a variety of relations can be derived. In particular it can be shown [**142**] that

$$\frac{j'}{j} = -\frac{E_6}{E_4}, \tag{VII.9}$$

$$\frac{j'}{j-1728} = -\frac{E_4^2}{E_6},$$

and

$$\frac{j''}{j'} = \frac{1}{6}E_2 - \frac{1}{2}\frac{E_4^2}{E_6} - \frac{2}{3}\frac{E_6}{E_4}. \tag{VII.10}$$

The following two formal power series, which can be shown to lie in $\mathbb{Z}[\zeta, 1/(\zeta(1-\zeta))][[q]]$, are crucial to the argument:

$$x(\zeta; q) = \frac{1}{12} - 2\sum_{n=1}^{\infty} \frac{q^n}{(1-q^n)^2} + \sum_{n\in\mathbb{Z}} \frac{\zeta q^n}{(1-\zeta q^n)^2}$$

and

$$y(\zeta; q) = \frac{1}{2}\sum_{n\in\mathbb{Z}} \frac{\zeta q^n(1+\zeta q^n)}{(1-\zeta q^n)^3}.$$

These power series can be shown to satisfy the following equality:

$$y^2 = x^3 - \frac{E_4(q)}{48}x + \frac{E_6(q)}{864}.$$

Projected to $\mathbb{F}_p$ (via reduction modulo $\mathfrak{B}$), the above equation means that, for the original elliptic curve equation $Y^2 = X^3 + aX + b$, we have

$$a = -\frac{\overline{E}_4(q)}{48}, \quad b = \frac{\overline{E}_6(q)}{864}. \tag{VII.11}$$

Back in $\mathbb{C}$, we also have the important relation

$$p_1 \triangleq \sum_{\zeta\in\mu_\ell,\ \zeta\neq 1} x(\zeta; q) = \frac{1}{12}\ell\left(E_2(q) - \ell E_2(q^\ell)\right), \tag{VII.12}$$

where μ_ℓ is the set of complex ℓth roots of unity. When projected to $\mathbb{F}_p$, this last expression will represent the sum of the x-components of points in the kernel of the isogeny, from which the coefficient of $x^{(\ell-3)/2}$ in the polynomial $F_\ell(x)$ will be directly derived. The relations established for the complex model will allow the computation of p_1, and the rest of the coefficients of $F_\ell(x)$, in $\mathbb{F}_p$.

To achieve this goal, we invoke, again, the modular polynomials. We start by deriving, in the next subsection, formulae based on the classical modular polynomials $\Phi_\ell(x, y)$. In a later subsection we briefly describe similar formulae for Müller's variant $G_\ell(x, y)$ [**110**], which are more computationally friendly.

VII.4.1. Using classical modular polynomials. It follows from the characterization of roots of $\Phi_\ell(x,j)$ in Formula (VII.7) that if $j(q)$ is the j-invariant of the original curve (interpreted over $\mathbb{C}$), and $\tilde{\jmath}\,(q) = j(q^\ell)$, then $\Phi_\ell(\tilde{\jmath}\,(q), j(q)) = 0$, i.e., given a curve with j-invariant $j(q)$, there exists an isogenous curve with j-invariant $\tilde{\jmath}\,(q) = j(q^\ell)$ and an isogeny of degree ℓ. Furthermore the following identities of power series can be established:

$$j'\Phi_x(j,\tilde{\jmath}\,) + \ell\tilde{\jmath}\,'\Phi_y(j,\tilde{\jmath}\,) = 0, \tag{VII.13}$$

and

$$\frac{j''}{j'} - \ell\frac{\tilde{\jmath}\,''}{\tilde{\jmath}\,'} = -\frac{j'^2\Phi_{xx}(j,\tilde{\jmath}\,) + 2\ell j'\tilde{\jmath}\,'\Phi_{xy}(j,\tilde{\jmath}\,) + \ell^2\tilde{\jmath}\,'^2\Phi_{yy}(j,\tilde{\jmath}\,)}{j'\Phi_x(j,\tilde{\jmath}\,)}, \tag{VII.14}$$

where the subscripts x and y denote partial derivatives with respect to those variables.

Equations (VII.13) and (VII.14) are of particular importance for the following development, where they will be interpreted over $\mathbb{F}_p$. Some care must be taken if some of the partial derivatives above vanish, as some of the relations obtained from the equations become void. The likelihood of this happening, however, is extremely low when working with random curves over very large fields. In case such a 'singularity' occurs, the random curve can be discarded, and another one selected. Therefore, we will ignore these cases. The problem is discussed in [**142**, p. 248].

To begin the computation, a value of $\tilde{\jmath} \in \mathbb{F}_p$ is required. This is found by considering $\gcd(x^p - x, \Phi_\ell(x,j))$. This GCD is usually a polynomial of degree two in this case, and one of its two roots is taken as $\tilde{\jmath}$. It can then be shown that the corresponding isogenous curve is given by

$$Y^2 = X^3 + \tilde{a}X + \tilde{b} \tag{VII.15}$$

where

$$\tilde{a} = -\frac{1}{48}\frac{(\tilde{\jmath}\,')^2}{\tilde{\jmath}\,(\tilde{\jmath}\, - 1728)}, \quad \tilde{b} = -\frac{1}{864}\frac{(\tilde{\jmath}\,')^3}{\tilde{\jmath}\,^2(\tilde{\jmath}\, - 1728)}, \tag{VII.16}$$

and where we have, from Equation (VII.13),

$$\tilde{\jmath}\,' = -\frac{j'\Phi_x(j,\tilde{\jmath}\,)}{\ell\Phi_y(j,\tilde{\jmath}\,)}. \tag{VII.17}$$

These computations take place in $\mathbb{F}_p$. For the original equation

$$Y^2 = X^3 + aX + b,$$

the relations in Equations (VII.11) define values of $\overline{E}_4(q), \overline{E}_6(q) \in \mathbb{F}_p$. For the isogenous curve in Equations (VII.15)–(VII.16), the similar relations

$$\tilde{a} = -\frac{\overline{E}_4(q^\ell)}{48}, \quad \tilde{b} = \frac{\overline{E}_6(q^\ell)}{864} \tag{VII.18}$$

define values of $\overline{E}_4(q^\ell)$, $\overline{E}_6(q^\ell) \in \mathbb{F}_p$.

It can further be shown that the sum of the x-coordinates in the kernel of the isogeny between these two curves corresponds precisely to the sum in Equation (VII.12), denoted by p_1 (which will denote the counterpart in $\mathbb{F}_p$). To apply the formula in Equation (VII.12), we also require counterparts of $E_2(q)$ and $E_2(q^\ell)$ in $\mathbb{F}_p$. These are obtained by using the relationship in Equation (VII.10), yielding, for p_1,

$$p_1 = \frac{\ell}{2}\left(\frac{j''}{j'} - \ell\frac{\tilde{j}''}{\tilde{j}'}\right) + \frac{\ell}{4}\left(\frac{\overline{E}_4^2(q)}{\overline{E}_6(q)} - \ell\frac{\overline{E}_4^2(q^\ell)}{\overline{E}_6(q^\ell)}\right) + \frac{\ell}{3}\left(\frac{\overline{E}_6(q)}{\overline{E}_4(q)} - \ell\frac{\overline{E}_6(q^\ell)}{\overline{E}_4(q^\ell)}\right). \qquad \text{(VII.19)}$$

The first term of the right-hand side of this equation is given by Equation (VII.14), where j' is obtained from Equation (VII.9). The remaining terms follow by direct computation.

Over $\mathbb{C}$, if the lattice corresponding to the curve is $\omega_1\mathbb{Z} + \omega_2\mathbb{Z}$, then the ℓ-isogeny under consideration is given by

$$\begin{array}{ccc} \mathbb{C}/(\omega_1\mathbb{Z}+\omega_2\mathbb{Z}) & \longrightarrow & \mathbb{C}/(\omega_1\mathbb{Z}+\ell\omega_2\mathbb{Z}) \\ z & \longmapsto & \ell z. \end{array}$$

Reducing everything modulo the prime ideal $\mathfrak{B}$ gives us the two ℓ-isogenous curves over $\mathbb{F}_p$ namely $Y^2 = X^3 + aX + b$ and $Y^2 = X^3 + \tilde{a}X + \tilde{b}$. In addition, the finite field isogeny is the reduction modulo $\mathfrak{B}$ of the complex isogeny.

Instead of the above isogeny, Schoof [**142**] finds it easier to work with the isogeny

$$\begin{array}{ccc} \mathbb{C}/(\omega_1\mathbb{Z}+\omega_2\mathbb{Z}) & \longrightarrow & \mathbb{C}/(\frac{1}{\ell}\omega_1\mathbb{Z}+\omega_2\mathbb{Z}) \\ z & \longmapsto & z \end{array}$$

for which the corresponding Weierstrass equation of the isogenous curve is

$$Y^2 = X^3 + \ell^4\tilde{a}X + \ell^6\tilde{b}.$$

Notice that this curve is isomorphic to the one with coefficients $\tilde{a}, \tilde{b}$, the two isogenies have the same kernel, and the preceding computation of p_1 is still correct.

Let $\wp(z)$ denote the Weierstrass function associated with the lattice L for the original curve, so

$$\wp(z) = \frac{1}{z^2} + \sum_{\omega\in L, \omega\neq 0}\left(\frac{1}{(z-\omega)^2} - \frac{1}{\omega^2}\right) = \frac{1}{z^2} + \sum_{k=1}^{\infty} c_k z^{2k} \qquad \text{(VII.20)}$$

where the coefficients c_k are obtained from the following recursion:

$$c_1 = -\frac{a}{5}, \quad c_2 = -\frac{b}{7}, \qquad \text{(VII.21)}$$

and

$$c_k = \frac{3}{(k-2)(2k+3)} \sum_{j=1}^{k-2} c_j c_{k-1-j}, \quad k \geq 3. \tag{VII.22}$$

The function $\tilde{\wp}$ for the isogenous curve is computed in a similar manner, using the curve coefficients $\ell^4\tilde{a}$ and $\ell^6\tilde{b}$. The analogous coefficients $\tilde{c}_k$ are then defined, using a recursion similar to that given in Equations (VII.21)–(VII.22).

The crucial observation, [**142**], is that if $F_\ell(x)$ is the polynomial with roots corresponding to the x-coordinates of the kernel of the isogeny,

$$\mathbb{C}/(\omega_1\mathbb{Z} + \omega_2\mathbb{Z}) \longrightarrow \mathbb{C}/(\frac{1}{\ell}\omega_1\mathbb{Z} + \omega_2\mathbb{Z})$$

then F_ℓ satisfies the equation

$$z^{\ell-1}F_\ell(\wp(z)) = \exp\left(-\frac{1}{2}p_1 z^2 - \sum_{k=1}^{\infty} \frac{\tilde{c}_k - \ell c_k}{(2k+1)(2k+2)} z^{2k+2}\right). \tag{VII.23}$$

Thus, from $a, b, \ell^4\tilde{a}$, and $\ell^6\tilde{b}$, we obtain the sequences c_k and $\tilde{c}_k$ using the recursion in Equations (VII.21)–(VII.22). From these sequences and p_1, in turn, the coefficients of $F_\ell(x)$ can be determined by expanding the functions on both sides of Equation (VII.23) and comparing like powers of z. Let $w = z^2$, and let $A(w)$ denote the function on the right-hand side of Equation (VII.23), expanded as a power series in w. Also, let $C(w) = \wp(z) - w^{-1} = \sum_{k=1}^{\infty} c_k w^k$, and, for an arbitrary power series $B(w)$, denote by $[B(w)]_j$ the coefficient of w^j in $B(w)$. If $F_\ell(x) = x^d + \sum_{i=0}^{d-1} F_{\ell,i}x^i$, then the coefficients of F_ℓ are given by the following recursion, where we set $F_{\ell,d} = 1$ and

$$F_{\ell,d-i} = [A(w)]_i - \sum_{k=1}^{i}\left(\sum_{j=0}^{k}\binom{d-i+k}{k-j}[C(w)^{k-j}]_j\right)F_{\ell,d-i+k}, \tag{VII.24}$$

for $1 \leq i \leq d$. Notice that at most d terms of each expansion are needed to determine the desired coefficients. Using the above recursion, the first few coefficients (from highest powers) of F_ℓ are given by

$$\begin{aligned}
F_{\ell,d-1} &= -\frac{p_1}{2}, \\
F_{\ell,d-2} &= \frac{p_1^2}{8} - \frac{\tilde{c}_1 - \ell c_1}{12} - \frac{\ell-1}{2}c_1, \\
F_{\ell,d-3} &= -\frac{p_1^3}{48} - \frac{\tilde{c}_2 - \ell c_2}{30} + p_1\frac{\tilde{c}_1 - \ell c_1}{24} - \frac{\ell-1}{2}c_2 + \frac{\ell-3}{4}c_1 p_1, \\
&\vdots
\end{aligned}$$

The calculation of the coefficients of F_ℓ over $\mathbb{F}_p$ requires that the denominators in the formulae above do not vanish. This can be guaranteed if p is sufficiently large that it exceeds the size of any factor of a denominator in the formulae. Noting that the c_k (or $\tilde{c}_k$) are required only for $k \leq d = O(\log p)$, that the largest factor of a denominator above associated with c_k is $2k+3$, and that the other denominators involve only small prime divisors of order $O(\ell)$ (as they all arise from factorials of numbers up to ℓ), we conclude that this condition is amply satisfied for the large fields $\mathbb{F}_p$ used in practice (after all, if p is small, no sophistication is needed to count points over $\mathbb{F}_p$). This requirement will be problematic, however, if an attempt is made to apply similar techniques to large finite fields of small characteristic. In Section VII.5 we describe Lercier's method to deal with fields of characteristic two. More general techniques for small characteristic are described by Couveignes [**33**], and are briefly discussed in Section VII.11.

The contents of the section are now summarized in algorithm form to indicate more directly how the factor $F_\ell(x)$ of the ℓth division polynomial $f_\ell(x)$ is computed.

ALGORITHM VII.3: **Division Polynomial Factor** $F_\ell(x)$

INPUT: An elliptic curve $E: y^2 = x^3+ax+b$ over $\mathbb{F}_p$ and an Elkies prime ℓ.
OUTPUT: A factor $F_\ell(x)$ of degree $d = \frac{\ell-1}{2}$ of $f_\ell(x)$.

1. Compute $j = j(E)$ from Equation (VII.8).
2. Compute $\overline{E}_4(q)$ and $\overline{E}_6(q)$ from Equations (VII.11).
3. Determine j' from Equation (VII.9).
4. Set $\tilde{j} \leftarrow$ a root of $\Phi_\ell(x, j)$ in $\mathbb{F}_p$.
5. Compute $\tilde{j}'$ from Equation (VII.17).
6. Compute $\tilde{a}$ and $\tilde{b}$ from Equations (VII.16).
7. Compute $\overline{E}_4(q^\ell)$ and $\overline{E}_6(q^\ell)$ from Equations (VII.18).
8. Compute $\frac{j''}{j'} - \ell\frac{\tilde{j}''}{\tilde{j}'}$ from Equation (VII.14).
9. Compute p_1 from Equation (VII.19).
10. Compute c_k and $\tilde{c}_k$ for $k \leq d$ from Equations (VII.21) and (VII.22).
11. Obtain the coefficients of $F_\ell(x)$ from the recursion in Equation (VII.24).
12. Return $F_\ell(x)$.

In Step 4, a root of $\Phi_\ell(x, j)$ is chosen as $\tilde{j}$, the j-invariant of the isogenous curve. In most cases, $\Phi_\ell(x, j)$ has two distinct roots in $\mathbb{F}_p$, and either choice will produce a correct F_ℓ. When $t^2 - 4p \equiv 0 \pmod{\ell}$, which can only happen if p is a square in $\mathbb{F}_\ell$, there may be either just one root, or $\ell+1$ roots in $\mathbb{F}_p$. In the latter case, again, any root may be chosen. On rare occasions, the procedure might fail to produce a factor F_ℓ, e.g., in cases where some denominator in the computation vanishes. In such cases, a different root of $\Phi_\ell(x, j)$ may be tried. If all roots fail in the same fashion, the trace modulo ℓ cannot be determined using this procedure (see [**142**] for a discussion of some of these singularities). However, the likelihood of this occurring (even for the first root), with random curves over very large finite fields, is extremely low. In any case, in a practical implementation, it is a good idea to check that the polynomial $F_\ell(x)$ produced by the algorithm is indeed a factor of the division polynomial $f_\ell(x)$.

Example. Consider the curve over $\mathbb{F}_{131}$ defined by

$$Y^2 = X^3 + X + 23,$$

and assume a factor of the division polynomial f_ℓ, with $\ell = 5$, is sought. All computations in the example are modulo 131. From the computations indicated in Steps 1–3 above, we obtain $j = 78$, $\overline{E}_4(q) = 83$, $\overline{E}_6(q) = 91$, and $j' = 66$. The modular polynomial $\Phi_5(x, y)$ from Section III.8, reduced modulo 131, and evaluated at $y = j = 78$, yields

$$\Phi_5(x, j) = x^6 + x^5 + 67x^4 + 106x^3 + 16x^2 + 33x + 41.$$

Its GCD with the field polynomial $x^{131} - x$ is

$$x^2 + 88x + 49 = (x - 17)(x - 26).$$

Thus, 5 is an Elkies prime for this curve. We try the root $\tilde{\jmath} = 17$. For this root, we obtain $\tilde{\jmath}\,' = 48$ from Equation (VII.17), where we computed the necessary derivatives of Φ_5 over $\mathbb{F}_{131}$. We next obtain $\tilde{a} = 62$, $\tilde{b} = 20$, and compute $\overline{E}_4(q^\ell) = 37$, $\overline{E}_6(q^\ell) = 119$. Next, we apply Equation (VII.14) to obtain

$$\frac{j''}{j'} - \ell\frac{\tilde{\jmath}\,''}{\tilde{\jmath}\,'} = 2,$$

and then, from Equation (VII.19), $p_1 = 42$. The coefficient $F_{5,1} = -p_1/2 = 110$ is then immediately derived. For this example, we only require the first term from each of the sequences c_k and $\tilde{c}_k$, namely, $c_1 = -a/5 = 26$, and $\tilde{c}_1 = -\tilde{a}\ell^4/5 = 110$. Finally, from the formula for the coefficient $F_{\ell,d-2}$, we obtain $F_{5,0} = 61$. Thus, we have $F_5(x) = x^2 + 110x + 61$, which is readily verified to be a factor of the division polynomial

$$\begin{aligned} f_5(x) &= x^{12} + 91x^{10} + 45x^9 + 110x^8 + 56x^7 + 93x^6 \\ &\quad + 21x^5 + 20x^4 + 36x^3 + 12x^2 + 16x + 103. \end{aligned}$$

Running the procedure with the second root, $\tilde{\jmath} = 26$, yields a different factor of $f_5(x)$, namely, $x^2 + 112x + 28$.

Analysis of the computation in Algorithm VII.3 reveals that its complexity is $O(\ell^3)$ operations in $\mathbb{F}_p$ (using naive arithmetic), or $O(\ell^2)$ (using fast methods) [**81**]. These estimates are within the complexity bounds of the steps of the SEA algorithm where F_ℓ is used. This makes the described construction of $F_\ell(x)$ a worthwhile computational investment that achieves the intended complexity gains over Schoof's original algorithm. This satisfactory assessment assumes, however, that the modular polynomials are available, and their coefficients have been reduced modulo p.

As mentioned in Section III.8, although the modular polynomials are used modulo primes p, their computation is done over $\mathbb{C}$, and the integers involved can grow extremely large, making the computation a daunting task. In addition, since presumably the point counting algorithm will be implemented to run with varying values of p, the polynomials are often stored in integer form. Therefore, although the complexity of Algorithm VII.3 is acceptable, for sufficiently large values of p the 'precomputation' step of obtaining the modular polynomials may be infeasible.

The situation can be significantly improved by using variants of the modular polynomials, whose coefficients do not grow as rapidly. One example is given by Müller's variant $G_\ell(x, y)$ of the modular polynomials [**110**], which was described in Section III.8. The derivation of F_ℓ based on these polynomials is described next. The emphasis is on the computational steps that differ slightly from those of this section, the underlying theory being quite similar. Notice that once the modular polynomials modulo p are available, the computational complexity of both methods is similar. Other alternatives for the modular polynomials are described, for instance, in [**108**] and [**40**].

VII.4.2. Using Müller's modular polynomials. We only derive the coefficient p_1 of the previous section. The other coefficients are derived in exactly the same manner as above.

As before, it is assumed that the coefficients a, b of an elliptic curve E defined over $\mathbb{F}_p$ are given, where p is a large prime. Also, all the following calculations are performed modulo p, even though the quantities involved are originally defined over $\mathbb{C}$. We just give the formulae, closely following Müller's thesis [**110**], where full explanations and proofs can be found.

We first compute a root, g, of the polynomial $G_\ell(x, j(E))$ given in Section III.8. Such a root must exist since we are assuming that ℓ is an Elkies prime. We set

$$\overline{E}_4 = -\frac{a}{3}, \; \overline{E}_6 = -\frac{b}{2}, \; \Delta = \frac{\overline{E}_4^3 - \overline{E}_6^2}{1728}.$$

We then compute, on setting $j = j(E)$,

$$\begin{aligned} D_g &= g\left(\frac{\partial}{\partial x} G_\ell(x, y)\right)(g, j), \\ D_j &= j\left(\frac{\partial}{\partial y} G_\ell(x, y)\right)(g, j), \end{aligned}$$

where the notation indicates the derivatives are to be evaluated at (g, j). The coefficients of the isogenous curve will be given by $\tilde{a}$ and $\tilde{b}$ and will have the associated invariants $\overline{E}_4^{(\ell)}$, $\overline{E}_6^{(\ell)}$, $\Delta^{(\ell)}$, etc. We can first deduce that

$$\Delta^{(\ell)} = \ell^{-12} \Delta g^{12/s},$$

where $s = 12/\gcd(l-1, 12)$.

If $D_j = 0$ then we are in a special case where $\overline{E}_4^{(\ell)} = \ell^{-2}\overline{E}_4$ and $\tilde{a} = -3\ell^4\overline{E}_4^{(\ell)}$. The j-invariant of the isogenous curve is given by $j^{(\ell)} = (\overline{E}_4^{(\ell)})^3/\Delta^{(\ell)}$ and $\tilde{b} = \pm 2\ell^6\sqrt{(j^{(\ell)} - 1728)\Delta^{(\ell)}}$. Finally in this special case we have $p_1 = 0$.

From now on we assume that $D_j \neq 0$, we then set

$$\overline{E}_2 = \frac{-12\overline{E}_6 D_j}{s\overline{E}_4 D_g}.$$

We then set

$$g' = -(s/12)\overline{E}_2^* g \ , \ j' = -\overline{E}_4^2\overline{E}_6\Delta^{-1} \ , \ \overline{E}_0 = \overline{E}_6(\overline{E}_4\overline{E}_2^*)^{-1},$$

where $\overline{E}_2^* = -12g'/sg$. Then we need to compute the quantities

$$\begin{aligned} D_g' &= g'\left(\frac{\partial}{\partial x}G_\ell(x,y)\right)(g,j) \\ &\quad + g\left[g'\left(\frac{\partial^2}{\partial x^2}G_\ell(x,y)\right)(g,j) + j'\left(\frac{\partial^2}{\partial x\partial y}G_\ell(x,y)\right)(g,j)\right], \\ D_j' &= j'\left(\frac{\partial}{\partial y}G_\ell(x,y)\right)(g,j) \\ &\quad + j\left[j'\left(\frac{\partial^2}{\partial y^2}G_\ell(x,y)\right)(g,j) + g'\left(\frac{\partial^2}{\partial y\partial x}G_\ell(x,y)\right)(g,j)\right], \end{aligned}$$

from which we can determine

$$\overline{E}_0' = \frac{1}{D_j}\left(\frac{-s}{12}D_g' - \overline{E}_0 D_j'\right).$$

We can then compute the value of $\overline{E}_4^{(\ell)}$, from

$$\overline{E}_4^{(\ell)} = \frac{1}{\ell^2}\left(\overline{E}_4 - \overline{E}_2^*\left[12\frac{\overline{E}_0'}{\overline{E}_0} + 6\frac{\overline{E}_4^2}{\overline{E}_6} - 4\frac{\overline{E}_6}{\overline{E}_4}\right] + \overline{E}_2^{*2}\right).$$

The j-invariant of the isogenous curve is then given by $j^{(\ell)} = \overline{E}_4^{(\ell)3}/\Delta^{(\ell)}$. We then need to compute the value of $\overline{E}_6^{(\ell)}$, which can be determined by setting $f = \ell^s g^{-1}$ and $f' = s\overline{E}_2^* f/12$, and then evaluating in turn the formulae,

$$\begin{aligned} D_g^* &= \left(\frac{\partial}{\partial x}G_\ell(x,y)\right)(f, j^{(\ell)}), \\ D_j^* &= \left(\frac{\partial}{\partial y}G_\ell(x,y)\right)(f, j^{(\ell)}), \\ j^{(\ell)\prime} &= -\frac{f' D_g^*}{\ell D_j^*}. \end{aligned}$$

We can now determine $\overline{E}_6^{(\ell)}$ from the equation

$$\overline{E}_6^{(\ell)} = -\frac{\overline{E}_4^{(\ell)} j^{(\ell)\prime}}{j^{(\ell)}}.$$

Finally we can compute our three desired quantities as

$$\begin{aligned}\tilde{a} &= -3\ell^4 \overline{E}_4^{(\ell)}, \\ \tilde{b} &= -2\ell^6 \overline{E}_6^{(\ell)}, \\ p_1 &= -\frac{\ell E_2^*}{2}.\end{aligned}$$

Notice that in this case the j-invariant of the isogenous curve could not be found as a root of the Müller variants of the modular polynomials. These polynomials were designed to have the same splitting type (with smaller coefficients) as the corresponding ordinary modular polynomial, but the roots of $G_\ell(x, j)$ do not correspond directly to j-invariants of isogenous curves.

The rest of the computation, to determine the remaining coefficients of $F_\ell(x)$, is the same as in the previous subsection, and is omitted here. Although the theory developed is intricate, as noted earlier in Section VII.2, it has been successfully used to establish the number of points on very large randomly chosen curves. Other references that pursue related approaches include [**40**] and [**26**].

Whether using classical modular polynomials or the variants of Müller, the factor $F_\ell(x)$ of the division polynomial has been determined for the case of large prime fields. The next section shows the technique developed by Lercier to achieve the same result for fields of characteristic two.

VII.5. Finding Factors of Division Polynomials through Isogenies: Characteristic Two

The work of Lercier on point counting for curves over fields of characteristic two is described in this section, using the references [**80**], [**85**] and [**82**].

As in the previous section, the goal is to find a factor $F_\ell(x)$ of degree $d = (\ell - 1)/2$ of the division polynomial $f_\ell(x)$, where ℓ is an Elkies prime. Here also, the problem will reduce to determining an isogenous curve, and then obtaining enough information about the kernel of the isogeny to produce the desired factor.

Attention is restricted to the non-singular curves of the form

$$E_{a_6} : Y^2 + XY = X^3 + a_6, \quad a_6 \in \mathbb{F}_{2^n}^*.$$

Recall the discriminant of this curve is a_6, and its j-invariant is $1/a_6$.

As in the odd characteristic case, we start by constructing the isogenous curve. This is done, as before, by finding a root $\tilde{j} \in \mathbb{F}_{2^n}$ of the modular polynomial $\Phi_\ell(x, j)$. In the characteristic two case, this leads immediately to the equation of the isogenous curve $E_{a_6'}$, as we have $a_6' = 1/\tilde{\jmath}$ in this case. From the knowledge of the two curves, sufficient information must be obtained about the points of the kernel to obtain $F_\ell(x)$.

A key result relating an isogeny with the points of its kernel is given by the application of Vélu's Theorem ([**160**], [**80**]) for fields of characteristic two. The theorem is a refinement of Theorem III.11, giving an explicit construction of the isogeny in terms of the kernel. Recall that for a subgroup R of the elliptic curve, we set $R^* = R \setminus \{\mathcal{O}\}$. As before, let P_X and P_Y denote, respectively, the x- and y-coordinates of a point P.

THEOREM VII.4. *Let R be a subgroup of odd order of an elliptic curve E_{a_6}. Define $a_6' = a_6 + \sum_{S \in R^*}(S_Y + (S_Y)^2)$. Then, there exist isogenies between E_{a_6} and $E_{a_6'}$, of kernel R. One such isogeny is*

$$\phi : P = (x, y) \mapsto \left(x + \sum_{S \in R^*} (P+S)_X,\ y + \sum_{S \in R^*} (P+S)_Y \right).$$

In our application, of course, the subgroup is not known and hence the isogeny cannot be derived directly in the manner of the theorem. On the other hand, a_6' is known, and Theorem VII.4 provides useful information that is exploited in the following theorem, which follows the formulation in [**82**].

THEOREM VII.5. *Let E_{a_6} and $E_{a_6'}$ be two isogenous elliptic curves defined over $\mathbb{F}_{2^n}$, such that the isogeny $\phi : E_{a_6} \to E_{a_6'}$ is of degree ℓ, an odd integer. Let $d = (\ell - 1)/2$. Then, ϕ can be expressed as*

$$\phi : (x, y) \mapsto \left(\frac{G(x)}{Q(x)^2}, \frac{H(x) + yK(x)}{Q(x)^3} \right)$$

where $Q(x)$, $G(x)$, $H(x)$ and $K(x)$ are in $\mathbb{F}_{2^n}[x]$ with degrees $d, 2d{+}1, 3d$ and $2d$ respectively. Furthermore, $G(x) = xP(x)^2$ where $P(x)$ is a polynomial of degree d such that $\gcd(P(x), Q(x)) = 1$ *and*

$$x^d Q(\sqrt{a_6}/x) = \frac{\sqrt[8]{a_6}}{\sqrt[8]{a_6'}} (\sqrt[4]{a_6})^d P(x),$$

or, by applying the change of variable $x \to \sqrt{a_6}/x$,

$$x^d P(\sqrt{a_6}/x) = \frac{\sqrt[8]{a_6'}}{\sqrt[8]{a_6}} (\sqrt[4]{a_6})^d Q(x).$$

To facilitate reference to the sources, we have preserved the notation of Lercier. Hence the polynomial $Q(x)$, of degree d in Theorem VII.5, will be equated to the sought factor $F_\ell(x)$ of the division polynomial $f_\ell(x)$.

The details of the proof can be found in [**80**] and [**85**]. It follows by applying Vélu's Theorem and the curve addition law. The kernel C of the isogeny is written as $\{\mathcal{O}\} \cup \mathfrak{S} \cup -\mathfrak{S}$, for a subset $\mathfrak{S}$ of size d, whose points exhaust all distinct x-coordinate values of points in C. Notice that, since ℓ

is odd, the point of order two is not in C, and thus $\mathfrak{S}$ and $-\mathfrak{S}$ are disjoint. Thus, using the addition law for points on the curve, an isogeny with the given kernel can be expressed as

$$(x,y) \stackrel{\phi}{\mapsto} \left(x\left(1+\sum_{S\in\mathfrak{S}} \frac{S_X}{(x-S_X)^2}\right), y+\sum_{S\in\mathfrak{S}}\left(\frac{y+x^2}{(x-S_X)^2}+\frac{x^2}{(x-S_X)^3}\right)S_X\right).$$

The first part of the theorem follows, after considerable detail, by letting $Q(x) = \prod_{S\in\mathfrak{S}}(x - S_X)$ (compare with Equation (VII.5)). It follows that $Q(x)$ divides the ℓth division polynomial, $f_\ell(x)$.

The second part of the theorem follows by observing also that $\phi \circ [2]_{a_6} = [2]_{a'_6} \circ \phi$, where the subscripts on the point multiplication maps indicate the curve they take place in. This observation also leads to the following corollary, which provides constraints that will eventually lead to an explicit construction of the polynomials $Q(x)$ and $P(x)$.

COROLLARY VII.6. *With the notation of the preceding theorem, the polynomials $P(x)$ and $Q(x)$ must satisfy the conditions*

$$x^d\hat{Q}(x+\sqrt{a_6}/x) = Q(x)P(x), \tag{VII.25}$$

and

$$(x+\sqrt[4]{a_6})\hat{P}(x+\sqrt{a_6}/x) = xP(x)^2+\sqrt[4]{a'_6}Q(x)^2, \tag{VII.26}$$

where $\hat{P}(x) = \sqrt{P(x^2)}$ and $\hat{Q}(x) = \sqrt{Q(x^2)}$, i.e. the polynomials whose coefficients are the square roots of those of $P(x)$ and $Q(x)$ respectively.

The following corollary follows from Theorem VII.5 and Corollary VII.6 (see [**80**] for a proof).

COROLLARY VII.7. *Let $P(x) = \sum_{i=0}^{d} p_i^2 x^i$, $Q(x) = \sum_{i=0}^{d} q_i^2 x^i$, $\alpha = \sqrt[4]{a_6}$ and $\beta = \sqrt[4]{a'_6}$. Then*

$$q_d = 1, \quad q_i = \frac{\sqrt[4]{\alpha}}{\sqrt[4]{\beta}}(\sqrt{\alpha})^{d-2i}p_{d-i}, \quad i \in \{0,1,\ldots,d\}, \tag{VII.27}$$

and

$$\left.\begin{array}{l} p_d = 1, \quad p_{d-1} = \alpha+\beta, \quad p_0 = \sqrt[4]{\alpha^{2d-1}\beta}, \\ p_{d-2} = \begin{cases} p_{d-1}^4 + \alpha p_{d-1} + \alpha^2 & \text{if } d \text{ is odd,} \\ p_{d-1}^4 + \alpha p_{d-1} & \text{if } d \text{ is even.} \end{cases} \end{array}\right\} \tag{VII.28}$$

Recall that the coefficients of the polynomial $P(x)$ (or, equivalently, $Q(x)$) are in $\mathbb{F}_{2^n}$ and that, by the preceding corollary, p_0, p_{d-2}, p_{d-1} and p_d are known. Comparing the coefficients that arise from an expansion of Equation (VII.26), and eliminating the q_i coefficients using Equation (VII.27), yields, for $k = 0, 1, \ldots, \lfloor \frac{d-1}{2} \rfloor$,

$$p_k^4 = \alpha^{2d-4k-1} \sum_{i=0}^{k} p_{d-2k-1+2i} B(d-2k-1+2i, i) \alpha^{2i} + \alpha^{2d-4k} \sum_{i=0}^{k} p_{d-2k+2i} B(d-2k+2i, i) \alpha^{2i}, \quad \text{(VII.29)}$$

and, for $k = 1, \ldots, \lfloor \frac{d}{2} \rfloor$,

$$p_{d-k}^4 = \alpha \sum_{i=0}^{k-1} p_{d+1-2k+2i} B(d+1-2k+2i, i) \alpha^{2i} + \sum_{i=0}^{k} p_{d-2k+2i} B(d-2k+2i, i) \alpha^{2i}, \quad \text{(VII.30)}$$

where $B(i,j)$ denotes the binomial coefficient $\binom{i}{j}$ (mod 2).

Several techniques to solve for $p_1, p_2, \ldots, p_{d-3}$ are proposed in [**80**] and [**85**]. Three of them are outlined next.

Method 1. Beginning with the known p_d, p_{d-1}, p_{d-2}, and p_0, from Equation (VII.29), one can expand the p_i in terms of a polynomial basis of $\mathbb{F}_{2^n}$ and treat the equations as a linear system of $n(d-2)$ equations in $n(d-3)$ binary variables (notice that the relations for $k = 0$ in Equation (VII.29) and $k = 1$ in Equation (VII.30) are already covered in Corollary VII.7). Unfortunately the complexity of this system, using conventional methods, is $O(\ell^3 n^3) = O(n^6)$ elementary $\mathbb{F}_2$ operations, which would defeat the purpose of seeking the factor F_ℓ.

Method 2. An improvement stems from the observation that, from Equation (VII.29), by setting $k = 1$ we get p_{d-3} as a function of p_1. From Equation (VII.30) with $k = 2$, in turn, we obtain p_{d-4} as a function of p_{d-3}. Iterating between the two Equations (VII.29) and (VII.30), the variables $p_{d-3}, \ldots, p_{i+1}$ can be expressed as functions of $p_1, \ldots, p_i$ for i approximately $d/3$. The remaining equations for $k \geq i$ contain variables raised to the power 2^j and thus are linear equations over $\mathbb{F}_2$, allowing reduction of a matrix one third the previous size (but still with the same asymptotics).

Method 3. An even more effective strategy results from casting the foregoing polynomial relations as a set of non-linear equations in Boolean indeterminates. To this end, the following $d+1$ additional relations are obtained by equating the coefficients of x^k, $k = 0, 1, \ldots, d$, on both sides of Equation (VII.25), and eliminating the q_i coefficients by means of Equation (VII.27):

$$\sqrt[4]{\alpha} \sum_{i=0}^{k} p_i^2 p_{d-k+i}^2 \alpha^{2i} = \sqrt[4]{\beta} \sqrt{\alpha}^{d+2k} \sum_{i=0}^{\lfloor k/2 \rfloor} p_{k-2i} B(d-k+2i, i). \quad \text{(VII.31)}$$

We first notice that the equations are quadratic in the unknowns p_i. Thus, for instance, starting from the equation for $k = 1$ (as the one for $k = 0$ just gives the value of p_0 from Corollary VII.7) and rearranging terms, we have

$$\sqrt[4]{\alpha^9}p_d^2p_1^2 + \sqrt[4]{\beta}(\sqrt{\alpha})^{d+2}p_1 + \sqrt[4]{\alpha}p_0^2p_{d-1}^2 = 0.$$

This is a quadratic equation in p_1, with coefficients expressed in terms of known quantities. After normalization, the equation can be rewritten as

$$p_1^2 + b_1p_1 + c_1 = 0.$$

Provided that $\mathrm{Tr}_{q|2}(c_1/b_1^2) = 0$, the equation has two solutions γ_1 and $b_1 + \gamma_1$, for some $\gamma_1 \in \mathbb{F}_{2^n}$ that can be computed explicitly using, for example, the methods of Section II.2.4. We write the solutions as $p_1 = \pi_0 b_1 + \gamma_1$, where π_0 is a (yet undetermined) Boolean variable. Since the curves E_{a_6} and $E_{a_6'}$ are known to be isogenous, the trace condition must always hold. With p_1 at hand (up to π_0), we proceed as in Method 2 above, turning to Equation (VII.29) with $k = 1$ to obtain p_{d-3} as a function of p_1, and thus of π_0. Similarly, we then turn to Equation (VII.30) with $k = 2$, and obtain p_{d-4} as a function of p_{d-4}, thus of π_0. Proceeding inductively, assume that after $K - 1$ iterations ($K \geq 2$) the coefficients $p_0, p_1, p_2, \cdots, p_{K-1}$, and $p_{d-2K}, p_{d-2K+1}, \cdots, p_d$ have been determined as functions of (undetermined) Boolean variables $\pi_0, \pi_1, \ldots, \pi_{K-2}$. These functions will take the form of multinomials in the π_i, with coefficients in $\mathbb{F}_{2^n}$, and where the degree in any of the π_i is at most one (since the π_i will eventually take on binary values, so $\pi_i^2 = \pi_i$). Coefficients p_i that are expressed as such multinomials will be said to be π-*determined*, the actual set of indeterminates π_i being understood from the context. At the Kth iteration, Equation (VII.31) with $k = K$ yields a quadratic equation

$$p_K^2 + b_Kp_K + c_K = 0,$$

with $b_K \in \mathbb{F}_{2^n}$, and c_K a polynomial function of previously π-determined coefficients p_i. Thus, c_K is π-determined. The quadratic equation has solutions expressed as

$$p_K = \pi_{K-1}b_K + \gamma_K,$$

where γ_K can be written explicitly as a function of b_K and c_K and is thus π-determined. This solution introduces a new Boolean variable, π_{K-1}. It is noted in [**80**], however, that the condition $\mathrm{Tr}_{q|2}(c_K/b_K^2) = 0$, which must hold if the quadratic equation has solutions, very often allows for eliminating an 'older' variable π_i, $i < K-1$. Thus, the number of 'active' Boolean variables grows rather slowly with K. In [**80**], this growth is heuristically estimated as logarithmic in K. After p_K is π-determined, Equation (VII.29) with $k = K$ and Equation (VII.30) with $k = K+1$ are used in turn to π-determine p_{d-2K-1} and p_{d-2K-2}, respectively. This part is similar to the iteration in Method 2.

This process continues for successive values of K until all coefficients p_i have been π-determined, which will occur when K is approximately $d/3$, depending on the value of d modulo 3. At that point, K relations remain unused from the sets in Equations (VII.29)–(VII.30). The π-determined p_i are substituted into these relations, resulting in a system of non-linear equations in the Boolean variables $\pi_0, \pi_1, \ldots, \pi_{K-2}$. Lercier describes a heuristic method for solving this system, which iterates on back-substitutions, with experimental evidence indicating that most variables are eventually isolated. When the number of remaining unsolved variables is small, exhaustive trial can be used to obtain a consistent solution. For more details of these heuristics, see [**80**] and [**85**].

A formal complexity analysis of the method is difficult, due to the ad hoc nature of some of the procedures involving the Boolean variables. Lercier [**80**] estimates the complexity at $O(\ell^3)$ operations in $\mathbb{F}_{2^n}$, based on heuristics and experimental evidence. This complexity would be within the bounds dictated by the other steps of the SEA algorithm, where F_ℓ is used. In any case, regardless of the lack of a formal proof of running time, the method has proven extremely effective in practice for counting points on elliptic curves over very large fields of characteristic two (e.g., $\mathbb{F}_{2^{1301}}$ reported in [**81**] and [**84**], or $\mathbb{F}_{2^{1663}}$ in a recent electronic communication by Lercier).

Solving for the coefficients p_i of $P(x)$ leads, through Corollary VII.7, to the determination of $Q(x)$. The latter, in turn, constitutes the desired factor $F_\ell(x)$ of the ℓth division polynomial $f_\ell(x)$.

The polynomial, $F_\ell(x)$, has now been determined for both the cases of large prime fields (Section VII.4) and characteristic two fields (this section). Thus, for either case, Step 5 of Algorithm VII.2 is complete.

VII.6. Determining the Trace Modulo a Prime Power

Beyond the computation of the trace t modulo a prime ℓ lies the possibility that it might be computed modulo a power of the prime, ℓ^k, for some positive integer k, hopefully without too much extra work. This problem has been considered, and it is of interest since it has the potential of reducing the magnitude of the largest prime required in the complete algorithm.

The somewhat surprising simplicity of the results in [**100**] is considered first, where the characteristic of the field is assumed to be two. Recall that in this case $E[2^c] \cong \mathbb{Z}/2^c\mathbb{Z}$ for any positive integer c.

LEMMA VII.8 (see [**100**]). *If* $\ell = 2^c$, $c \geq 2$, *then* $f_\ell(x)$ *has a factor* $f(x)$ *of degree* $\ell/4$ *in* $\mathbb{F}_{2^n}[x]$.

The proof is immediate. Since $E[\ell] \cong \mathbb{Z}/2^c\mathbb{Z}$, $f_\ell(x)$ has only $\ell/2$ distinct roots (taking into account the point at infinity and the unique point of order two, $(0, \sqrt{a_6})$) and of these, only $\ell/4$ are x-coordinates of points of order ℓ. Since the set of such points is stable under the Frobenius map, the ℓth division polynomial has a factor of degree $\ell/4$. Remarkably, the next lemma shows how these polynomials can be constructed recursively.

LEMMA VII.9 (see [**100**]). *Let $\ell = 2^c$ and define the sequence of polynomials $\{g_i(x)\}$ in $\mathbb{F}_{2^n}[x]$ as follows:*

$$g_0(x) = x,$$

$$g_1(x) = b_1 + x \qquad \text{where } a_6 = b_1^4,$$

$$g_i(x) = (g_{i-1}(x))^2 + b_i x \prod_{j=1}^{i-2} (g_j(x))^2 \quad \text{where } a_6 = b_i^{2^{i+1}},\ i \geq 2.$$

Then $g_{c-1}(x)$ is a factor of degree $\ell/4$ of $f_\ell(x)$ in $\mathbb{F}_{2^m}[x]$. Moreover, the roots of $g_{c-1}(x)$ are precisely the x-coordinates of points of order ℓ.

Of course, the higher the power of two that can be used in the trace computation, the fewer other primes need be used. On the other hand, too high a power of two would increase the complexity of computations involving $g_{c-1}(x)$ above. In practice, a good choice is $c \approx \log_2 n$, so that $\ell = 2^c$ is of the same order of magnitude as the largest primes used in the algorithm.

More generally, for using powers of primes, the work of Couveignes [**35**] and of Couveignes and Morain [**36**] is noted. The idea in this work is to create a cycle of isogenies between curves (finite since there is a finite number of such curves). For an Elkies prime ℓ one finds the two roots of $\Phi_\ell(x, j)$ where j is the j-invariant of the given curve. Choosing one of these, j_1, one finds next a root of $\Phi_\ell(x, j_1)$, and so on. The process leads to a polynomial of degree $\ell^{k-1}(\ell-1)/2$, which one uses to compute t modulo ℓ^k. Notice that in the case of characteristic 2, and $\ell = 2^k$, this polynomial is of degree $\ell^{k-1}/2 = 2^{k-2} = \ell/4$. The reader is referred to the references [**35**] and [**36**] for the details.

VII.7. The Elkies Procedure

Sections VII.4 and VII.5 have described the construction of the polynomial $F_\ell(x)$, a factor of degree $(\ell-1)/2$ of the ℓth division polynomial, for the Elkies prime ℓ, in the cases of odd and even characteristic respectively. Section VII.6 discussed the construction of a similar polynomial, of even lower degree, for the case where ℓ is a power of a small prime. In this section we review how F_ℓ is used to determine the trace of Frobenius modulo ℓ. The technique was already mentioned in Section VII.2.1, so only a brief description is given here.

The computation is very similar to that of the basic Schoof algorithm of Section VII.1, with the two exceptions of using the polynomials $F_\ell(x)$ in the place of the division polynomials $f_\ell(x)$ and the fact that, for an Elkies prime ℓ, we are guaranteed the existence of eigenspaces of the Frobenius map in the group of torsion points.

Suppose the Frobenius endomorphism φ has the eigenvalue $\lambda \in \mathbb{F}_\ell$ on C, the Galois stable subgroup of $E[\ell]$ defined by the roots of F_ℓ. This means that, for $(x, y) \in C$,

$$\varphi(x,y) = (x^q, y^q) = [\lambda](x,y). \tag{VII.32}$$

Since φ satisfies the equation

$$\varphi^2 - [t]\varphi + [q] = [0]$$

we have $t \equiv \lambda + q/\lambda \pmod{\ell}$. Hence to compute $t \pmod{\ell}$ in this case, it is sufficient to find the eigenvalue λ satisfying Equation (VII.32). As before, this can be done by checking for values $1 \le \lambda \le \ell - 1$, and, since ℓ is prime, one and only one such value will satisfy the equation. It is noted, however, that not all values of λ need be checked since, much as for the Atkin case, Case (ii) of Proposition VII.2 also restricts the number of possible values, in addition to the reduction stemming from having to check only one value from each pair $\pm\tau$.

To check Equation (VII.32), it is necessary to compute x^q and y^q modulo the polynomial $F_\ell(x)$ and the curve equation. Using an idea similar to that used in the original Schoof algorithm, one computes, in the case of large prime characteristic,

$$\begin{aligned} h(x,y) &= \left((x^p - x)\psi_\lambda^2(x,y) + \psi_{\lambda-1}(x,y)\psi_{\lambda+1}(x,y)\right) \\ &\qquad\qquad\qquad\qquad \pmod{F_\ell(x), y^2 - x^3 - ax - b} \\ &= a(x) + yb(x), \end{aligned}$$

where in this case ψ_i are the bi-variate division polynomials. As before, one checks the existence of a point in C that satisfies this equation by computing $H(x) = \gcd(a(x), b(x), F_\ell(x))$. If $H(x) = 1$ then λ is not an eigenvalue and if $H(x) \neq 1$ then the corresponding y-coordinates are checked. The details for fields of characteristic 2 are similar.

At this point, Algorithm VII.2 has been completed down to Step 8.

VII.8. The Atkin Procedure

In what follows, it is assumed that ℓ is an Atkin prime, i.e., $t^2 - 4q$ is not a square in $\mathbb{F}_\ell$. A slightly more detailed version of the Atkin algorithm than that outlined in Section VII.2.2 is given here.

The previous information can be interpreted as follows. If t is the trace of the Frobenius endomorphism so that $\varphi^2 - [t]\varphi + [q]$ is the zero map, and if ℓ is an Atkin prime, then the polynomial $x^2 - tx + q \pmod{\ell}$ splits in $\mathbb{F}_{\ell^2}$, with two roots λ and μ such that $\lambda, \mu \in \mathbb{F}_{\ell^2} - \mathbb{F}_\ell$. The element $\lambda/\mu = \gamma_r$ is then an element of order exactly r in $\mathbb{F}_{\ell^2}$ and it comes from a set of order $\phi_{\mathrm{Eul}}(r)$. As noted, this is also the number of possible values of the trace in this case.

The elements of order r in $\mathbb{F}_{\ell^2}$ can be determined easily by first finding a generator g for $\mathbb{F}_{\ell^2}^*$, and then computing $\gamma_r = g^{i(\ell^2-1)/r}$, where $i \in \{1, \ldots, r-1\}$ is coprime to r.

Enumerating all possibilities for γ_r we obtain a set of possible values for $t \pmod{\ell}$. These are obtained from the equations

$$t = \lambda + \mu \pmod{\ell}, \quad q = \lambda\mu \pmod{\ell}, \quad \text{and } \gamma_r = \lambda/\mu.$$

To show the simplicity of the technique, the steps will be explained in detail. Write $\mathbb{F}_{\ell^2} = \mathbb{F}_\ell[\sqrt{d}]$, $\lambda = x_1 + \sqrt{d}x_2$ and $\gamma_r = g_1 + \sqrt{d}g_2$, for a quadratic non-residue $d \in \mathbb{F}_\ell$. The values of x_1 and x_2 are not known, but the possible values for g_1 and g_2 are. Since μ is the conjugate of λ we have $\mu = x_1 - \sqrt{d}x_2$, from which the following equation is derived:

$$\begin{aligned} g_1 + \sqrt{d}g_2 &= \gamma_r = \frac{\lambda}{\mu} = \frac{\lambda^2}{\lambda\mu} \\ &= \frac{1}{q}\left(x_1^2 + dx_2^2 + 2x_1x_2\sqrt{d}\right). \end{aligned}$$

Hence

$$\begin{aligned} qg_1 &\equiv x_1^2 + dx_2^2 \pmod{\ell}, \\ qg_2 &\equiv 2x_1x_2 \pmod{\ell}, \\ q &\equiv x_1^2 - dx_2^2 \pmod{\ell}. \end{aligned}$$

Hence, $x_1^2 = q(g_1+1)/2$, from which at most two possible values for x_1 can be derived. The required value of $t \pmod{\ell}$ is then obtained from $t \equiv 2x_1 \pmod{\ell}$. An expansion of the Atkin section of Algorithm VII.2 (Steps 10–11) is given below, for a fixed Atkin prime ℓ. as follows:

ALGORITHM VII.4: Atkin Procedure.

```
INPUT:  A curve E over a finite field F_q and a prime ℓ.
OUTPUT: A pair (T,ℓ), T the set of the possible traces t (mod ℓ).
1.  T←{}.
2.  Determine the splitting behaviour of Φ_ℓ(x,j) in F_q.
3.  Determine r using Proposition VII.2.
4.  Determine a generator g of F*_{ℓ^2} = F_ℓ[√d]*.
5.  S←{g^{i(ℓ^2-1)/r} : (i,r) = 1}.
6.  For each γ_r ∈ S do:
7.      Write γ_r = g_1 + √d g_2.
8.      z←q(g_1+1)/2 (mod ℓ).
9.      If z is a square modulo ℓ then do:
10.         x←√z (mod ℓ).
11.         T←T ∪ {2x,-2x}.
12. Return (T,ℓ).
```

For each Atkin prime ℓ, a set $T = \{t_1, \ldots, t_{\phi_{\mathrm{Eul}}(r)}\}$ of possible traces modulo ℓ is obtained. In many cases r is a relatively small integer making the search described later for the exact value simpler. However, the complexity-conscious reader will have undoubtedly noticed that, even if the sets T are relatively small, the number of possible values of the trace one needs to check grows exponentially with the number of Atkin primes. This number, in turn, could be about one half the number of primes considered, or $O(\log q)$. This means that if the algorithm processes all the Atkin primes it encounters, the complexity is, in effect, exponential in $\log q$.

A way out of this problem, from a complexity-theoretic point of view, is to just use Elkies primes. However, this implies having to deal with modular polynomials of higher degree, which is in itself a problem. So the best practical compromise is obtained by judicious use of the Atkin procedure, where only the 'best' Atkin primes (i.e. those giving small sets T) are retained, and the overall size of the set of potential traces is bounded. In addition, the search for the exact value of the trace among the candidates defined by the Atkin algorithm can be significantly accelerated by means of a BSGS procedure described in the next section. Thus, for practical values of $\log q$, the Atkin procedure still plays a useful role in the algorithm, although as $\log q$ increases, the proportion of Elkies primes used needs to increase, to maintain a computational complexity balance. An example of this trade off is given at the end of Section VII.10, for a fairly large value of $\log q$.

Next, we show how to combine the Atkin information with the exact values obtained from the Elkies primes, to determine the exact value of the trace of Frobenius, and hence the group order of the curve.

VII.9. Combining the Information from Elkies and Atkin Primes

At this point we have completed down to Step 13 of Algorithm VII.2, assuming a sufficiently large number of primes has been considered to satisfy Step 2. It remains to complete the algorithm by finding the exact value of t from the information gathered. Our exposition will follow that given in Müller's thesis [**110**].

The data from the Elkies primes is combined via the CRT, to determine two numbers t_3 and m_3 such that

$$t \equiv t_3 \pmod{m_3}.$$

Here, m_3 is the product of all the Elkies primes used. The set of Atkin primes is divided into two sets such that each set gives roughly the same number of possible traces modulo the respective moduli. Again using the CRT on these two sets in turn, we determine two moduli m_1 and m_2 and two sets S_1 and S_2 such that

$$\begin{aligned} t &\equiv t_1 \pmod{m_1} \text{ with } t_1 \in S_1, \\ t &\equiv t_2 \pmod{m_2} \text{ with } t_2 \in S_2. \end{aligned}$$

Clearly m_1, m_2 and m_3 are pairwise coprime. It will be assumed that enough Atkin and Elkies primes have been taken so that

$$m_1 m_2 m_3 > 4\sqrt{q}.$$

Since $|t| \leq 2\sqrt{q}$, if $t \pmod{m_1 m_2 m_3}$ is determined then we will have found t exactly and hence the group order.

The exact value of t is now determined by the type of space/time trade off seen in the BSGS. Note that we can write

$$t = t_3 + m_3(m_1 r_2 + m_2 r_1)$$

for some integers r_1 and r_2 with

$$\begin{aligned} r_1 &\equiv \frac{t_1 - t_3}{m_2 m_3} \pmod{m_1}, \\ r_2 &\equiv \frac{t_2 - t_3}{m_1 m_3} \pmod{m_2}, \end{aligned}$$

where $t \equiv t_1 \pmod{m_1}$ and $t \equiv t_2 \pmod{m_2}$. But the exact values of t_1 and t_2 are not known; all that is known is that they come from the finite sets S_1 and S_2.

Although the above formulae give r_1 and r_2 modulo m_1 and m_2, they say nothing whatever about the required sizes of r_1 and r_2. Fortunately the following lemma is available:

LEMMA VII.10. *If we choose*

$$0 \leq t_3 < m_3 \text{ and } \lfloor \frac{-m_1}{2} \rfloor < r_1 \leq \lfloor \frac{m_1}{2} \rfloor,$$

then $|r_2| \leq m_2$.

PROOF. Using the above equation it is seen that

$$r_2 = \frac{1}{m_1 m_3} (t - t_3 - m_2 m_3 r_1).$$

Hence,

$$\begin{aligned} |r_2| &\leq \frac{1}{m_1 m_3} (|t| + |t_3| + m_2 m_3 |r_1|) \\ &\leq \frac{2\sqrt{q}}{m_1 m_3} + \frac{1}{m_1} + \frac{m_2}{2} \\ &\leq \frac{m_2}{2} + \frac{1}{m_1} + \frac{m_2}{2}, \end{aligned}$$

since $m_1 m_2 m_3 > 4\sqrt{q}$. □

The group order of our curve is $q + 1 - t$, and so for any point $P \in E(\mathbb{F}_q)$ we must have

$$[q+1]P = [t]P = [t_3 + m_3(m_1 r_2 + m_2 r_1)]P.$$

Rearranging this a little yields

$$[q+1-t_3]P - [r_1 m_2 m_3]P = [r_2 m_1 m_3]P.$$

It should now be clear how to proceed to determine r_1 and r_2, and hence t. A random point, P, is chosen on the curve, which does not have an obviously small order. For every possible value of $t_1 \pmod{m_1}$ the corresponding value of $r_1 \equiv (t_1 - t_3)/m_2 m_3 \pmod{m_1}$ is computed. Taking the value of r_1 such that $|r_1| \le \lfloor m_1/2 \rfloor$, compute

$$Q_{r_1} = [q+1-t_3]P - [r_1 m_2 m_3]P$$

and store the value (Q_{r_1}, r_1) in a table which is sorted on the Q_{r_1}. This table is sorted so as to allow for easy table lookup in the following phase. This can be thought of as the table creation phase of the baby steps in BSGS.

We now proceed with the analogue of the giant steps. Each possible value of t_2 is taken in turn and the corresponding value of $r_2 \equiv (t_2 - t_3)/m_1 m_3 \pmod{m_2}$ is computed. We will need to take all such values of r_2 in the set $\{-m_2, \dots, m_2\}$. The points

$$R_{t_2} = [r_2 m_1 m_3]P$$

are computed and it is checked whether R_{t_2} is one of the elements in the table computed earlier. If so, then we have determined a pair of 'matching' r_1 and r_2. This pair allows the determination of a possible value for t, which in turn gives a possible value for the group order, m.

This group order can then be checked to be correct by means of random curve points, as discussed in Section VI.2.

VII.10. Examples

Example 1. The number of points on the elliptic curve

$$E : Y^2 + XY = X^3 + 1$$

over the finite field $\mathbb{F}_{2^6}$ is computed. This is a contrived example, since not only is it quite small but in addition the number of points could be computed from a recurrence relation, once the number of points in $E(\mathbb{F}_2)$ (which is, trivially, 4) has been determined. It is nonetheless instructive.

First, notice that only the trace modulo a number larger than $4\sqrt{64} = 32$ need be determined. Hence the trace modulo the primes 2, 3, 5 and 7 is found.

$\ell = 2$. Since the curve is not supersingular we know that the trace is an odd number. Hence $t \equiv 1 \pmod 2$.

$\ell = 3$. The prime 3 is an Atkin prime, since the modular polynomial $\Phi_3(s,t)$ factors over $\mathbb{F}_{2^6}$ as a product of two quadratic polynomials. The method for Atkin primes with $r = 2$ is then applied. Hence there is only one possibility for the trace modulo 3 and that is 0.

$\ell = 5$. The prime 5 is an Elkies prime and we can quickly determine that 2 is an eigenvalue of the Frobenius morphism modulo 5. Hence the trace modulo 5 is equal to 4.

$\ell = 7$. This case is also an Elkies prime, with eigenvalue 1. The trace is therefore 2 modulo 7.

Using the CRT it is found that 9 is the unique integer, t, modulo $210 = 2{\cdot}3{\cdot}5{\cdot}7$ for which

$$\begin{aligned} t &\equiv 1 \pmod 2), \\ t &\equiv 0 \pmod 3), \\ t &\equiv 4 \pmod 5), \\ t &\equiv 2 \pmod 7). \end{aligned}$$

Hence the actual trace, which must satisfy $|t| \leq 2\sqrt{q} = 16$, is also equal to 9. But this means the group order is equal to $q + 1 - t = 64 + 1 - 9 = 56$. This group order can then be verified either by using the recurrence sequence mentioned earlier or by multiplying a set of random points on $E(\mathbb{F}_{2^6})$ by 56.

Example 2. As a more challenging example consider the curve

$$E : Y^2 + XY = X^3 + 1 + \theta^{18}$$

over $\mathbb{F}_{2^{20}}$ where

$$\theta^{20} + \theta^3 + 1 = 0.$$

The four odd primes less than 13 are used and it is found that there are two Elkies and two Atkin primes. The data for the Elkies primes is

Prime	Eigenvalue	Trace
7	1	5
11	4	7

The data for the Atkin primes is summarized in the table

Prime	r-Value	Possible Traces
3	2	0
5	3	1 or 4

Using the method for determining the trace modulo 2^ℓ, it is found that the trace modulo 8 is given by 1.

Putting all this information together, and testing the possible traces with various random points on $E(\mathbb{F}_{2^{20}})$, it is seen that the trace is equal to 1041. Hence the group order is equal to 1047536. Such an example as this takes well under a second to determine the group order, whilst a brute force enumeration of the group order takes over ten seconds.

Example 3. This is a more challenging example, which is still too small for use in cryptography. We take the curve

$$E : Y^2 = X^3 + 1150871587567420791922262222331X + 5413109024187593793299830671 19$$

over the field of $p = 1267650600228229401496703205653 = 2^{100} + 277$ elements. Using all the primes up to 41 we find that there are four Elkies primes and ten Atkin primes. The data we summarize below; first the Elkies primes,

Prime	Eigenvalue	Trace
13	−6	8
19	−2	3
29	−13	24
43	18	36

and now the Atkin primes,

Prime	Possible Traces
2	1
3	1 or 2
5	0
7	1, 3, 4 or 6
11	5 or 6
17	3, 6, 8, 9, 11 or 14
23	2, 10, 13 or 21
31	3, 5, 7, 8, 9, 10, 12, 15, 16, 19, 21, 22, 23, 24, 26 or 28
37	1, 2, 8, 9, 10, 12, 14, 15, 18, 19, 22, 23, 25, 27, 28, 29, 35 or 36
41	9, 10, 15, 26, 31 or 32

Putting all this data together we find that the group order is equal to

$$1267650600228229462216521077879$$

which is

$$13 \cdot 97511584632940727862809313683.$$

Example 4. Here, we consider a 'real life' curve over a very large field of characteristic two, $\mathbb{F}_{2^{431}}$. In fact, recalling the key length comparisons of Section I.3, this field size probably exceeds the security requirements of most cryptographic applications. The curve parameters and order are listed in Example 23 in the Appendix. The SEA algorithm, using Lercier's improvements, was run on the twist of the curve.

First, the value of the trace modulo 64 was determined using the techniques of Section VII.6. Then, prime values of $\ell \geq 3$ were considered. Of them, the following 22 numbers were found to be Elkies primes, and the trace of Frobenius modulo each of them was determined:

$$13, 19, 41, 43, 59, 61, 71, 73, 97, 103, 109, 113, 127,$$
$$131, 139, 167, 173, 179, 181, 193, 197, 199.$$

The product of 64 and the 22 Elkies primes is a 150-bit integer. The number of bits of uncertainty for the trace of Frobenius is 218. The remaining gap of 68 bits is closed with the aid of the Atkin primes.

The following 24 numbers were found to be Atkin primes:

$$3, 5, 7, 11, 17, 23, 29, 31, 37, 47, 53, 67, 79, 83, 89,$$
$$137, 151, 191, 211.$$

Due to the complexity considerations discussed in Section VII.8, only a subset consisting of 13 of these primes, and the associated possible trace values, were kept. Those kept were divided into two sets, one for the 'baby' steps, one for the 'giant' steps, as discussed in Section VII.9. The two sets are listed below. For each prime, the number of potential trace values modulo that prime is listed. Also, the total number of traces for each set is listed.

Baby steps

Prime	# Traces
5	2
11	4
23	4
31	1
47	4
151	18
211	2
Total traces	4608

Giant steps

Prime	# Traces
7	2
29	1
53	18
79	4
89	8
191	8
Total traces	9216

The product of the Atkin primes kept is a 70-bit integer, closing the gap as claimed. As seen in the example, even though the Atkin procedure is in principle of exponential complexity, the BSGS technique and judicious management of the traces kept allow for good utilization of Atkin primes within very reasonable complexity limits.

VII.11. Further Discussion

It should be noted that there are two other algorithms for the computation of isogenies of degree ℓ, both due to Couveignes ([**34**], [**33**]) and both valid over any field characteristic. The first uses the theory of formal groups of elliptic curves to establish the isogeny of degree ℓ, and thence the factor of the ℓth division polynomial, $F_\ell(X)$. This algorithm is characterized in [**33**] as being of the same asymptotic complexity as Lercier's, but in practice slower by a significant constant factor. The second Couveignes algorithm notes that for a field of characteristic p, the p^kth division polynomial is a p^kth power of a polynomial $\tilde{f}_{p^k}$ of degree $d_k = (p^k - 1)/2$ if p is odd and $2^k - 1$ if p is even. Suppose an isogeny ϕ of degree ℓ, $\gcd(\ell, p) = 1$, maps the curve E_1 to E_2. Let P_1 be a primitive point of $E[p^k]$ in E_1. Suppose the isogeny maps this point to a primitive point P_2 in $E_2[p^k]$. Then the isogeny maps $[m]P_1$ to $[m]P_2$ for all $1 \leq m < p^k$. Let $A(X)$ be the polynomial of degree $d_k - 1$ such that $A(([m]P_1)_X) = ([m]P_2)_X$, $1 \leq m \leq d_k$. It follows that

$$\phi(X) = \frac{G_\ell(X)}{F_\ell^2(X)} \equiv A(X) \pmod{\tilde{f}_{p^k}(X)}.$$

Knowing the polynomial $A(X)$ and the division polynomial factor $\tilde{f}_{p^k}(X)$, this equation can be solved for $G_\ell(X)$ and $F_\ell(X)$, the factor of the division polynomial sought, by either the Berlekamp [**9**] or extended Euclidean [**61**] algorithms. Of course, $E[p^k]$ may not be rational and the interpolation to find $A(X)$ must be done over an extension field. In addition it is required to find primitive p^k-torsion points. Further details on these and other aspects of the algorithm can be found in [**84**] and [**33**].

CHAPTER VIII

Generating Curves using Complex Multiplication

The CM method of computing elliptic curves over a finite field uses the theory of complex multiplication of elliptic curves over the rationals. The method for finding elliptic curves over a field of large prime characteristic will be described in this chapter. The case of characteristic two follows with some minor modification, and the details can be found in [**73**], although a brief outline will be given below. The main ideas for the large prime characteristic case originally arose in the context of the elliptic curve based primality proving algorithm (see [**7**] and Chapter IX). An accessible account of the method can be found in [**107**].

VIII.1. The Theory of Complex Multiplication

Only a brief outline of the theory will be given here. The interested reader should consult a book, such as [**148**] or [**29**], for the details. As remarked earlier two elliptic curves are isomorphic over the algebraic closure if their j-invariants are equal. In discussing j-invariants two special cases, given by $j = 0$ and $j = 1728$, need to be singled out.

Given a complex j-invariant an elliptic curve over $\mathbb{C}$, with this j-invariant, can be written down using the following rule:

- If $j = 0$ then use
$$E : Y^2 = X^3 - 1.$$
- If $j = 1728$ then use
$$E : Y^2 = X^3 - X.$$
- Otherwise set $c = j/(j - 1728)$ and use
$$E : Y^2 = X^3 - 3cX + 2c.$$

Compare this with Lemma VIII.3 below.

Consider the ring of endomorphisms, $\mathrm{End}(E)$, of a curve, E, defined over an arbitrary field K. As mentioned in Chapter III, if E is not supersingular, $\mathrm{End}(E)$ either is equal to $\mathbb{Z}$ or is equal to an order in an imaginary quadratic number field. If $\mathrm{End}(E)$ is equal to an order in an imaginary quadratic number field then the curve is said to have CM. In such a situation

$$\mathrm{End}(E) \cong \mathbb{Z} + \mathbb{Z}\tau$$

where τ is a complex algebraic number of degree two, such that $\tau \in \mathcal{H}$, where $\mathcal{H}$ is the so-called Poincaré half-plane, the upper half (positive imaginary part) of the complex plane.

The theory of complex multiplication and j-invariants are linked, at least for the moment in the case of characteristic zero, via the result

THEOREM VIII.1. *Suppose $\tau \in \mathcal{H}$, with τ a complex algebraic number of degree two. Then when we set $E_\tau = \mathbb{C}/(\mathbb{Z} + \mathbb{Z}\tau)$ we have*

(1) $\mathrm{End}(E_\tau)$ *is an order in* $\mathbb{Q}(\tau)$, *hence* E_τ *has complex multiplication,*
(2) $j(\tau) = j(E_\tau)$ *is an algebraic integer.*

It is the latter of these properties which shall be exploited below. Notice how surprising it seems at first sight; $j(\tau)$ is defined by the Fourier series in Chapter III, so why should its value at a complex quadratic number be an algebraic integer?

The main theorem is the following.

THEOREM VIII.2. *Let $\tau \in \mathcal{H}$ be a complex quadratic number with discriminant $-D$. Hence $-D$ is the discriminant of the primitive positive definite quadratic form $Q(x,y)$ which has τ as a root of $Q(x,1) = 0$. Let h_D denote the class number of the order of discriminant $-D$. Then $j(\tau)$ is an algebraic number of degree h_D and its minimal polynomial is given by*

$$H_D(x) = \prod (x - j(\alpha))$$

where α runs over all complex numbers such that $(\alpha, 1)$ is a zero of one of the h_D inequivalent primitive reduced forms of discriminant $-D$.

If $\mathbb{Z}[\tau]$ is the maximal order of some imaginary quadratic number field, K, then

$$H = K(j(\tau))$$

is an extension of K of degree h_D. In fact it is the maximal unramified abelian extension of K. Since H is Galois over K we can consider its Galois group. The Galois group of H over K is isomorphic to the class group of K. By definition, H is called the Hilbert class field of K; it is a field under which every ideal in $\mathbb{Z}[\tau]$ becomes principal when considered as an ideal in $\mathbb{Z}_H$.

In the rest of this chapter it will be assumed, as above, that $\mathbb{Z}[\tau]$ is the maximal order of some imaginary quadratic number field and so $-D$ will represent a fundamental discriminant. Hence $-D$ is a number congruent to 1 or 0 modulo 4 and no odd prime divides D to a power greater than one. We shall refer to d as the square free positive integer such that $\mathbb{Q}(\tau) = \mathbb{Q}(\sqrt{-d})$, in other words if $d \equiv 3 \pmod 4$ then $D = d$, otherwise $D = 4d$.

It will be required to compute the Hilbert class polynomial, $H_D(x)$, of the preceding theorem. In particular we will need to compute, to high precision, the values of $j(\tau)$ for various $\tau \in \mathcal{H}$. As noted in Chapter III we can compute

$j(\tau)$ efficiently using the formulae

$$h(\tau) = \frac{\Delta(2\tau)}{\Delta(\tau)}\ ,\ j(\tau) = \frac{(256h(\tau)+1)^3}{h(\tau)},$$

where $\Delta(\tau)$ is computed using

$$\Delta(\tau) = q\left(1 + \sum_{n\geq 1}(-1)^n\left(q^{n(3n-1)/2} + q^{n(3n+1)/2}\right)\right)^{24},$$

where $q = \exp(2\pi\sqrt{-1}\tau)$. Thus $H_D(x)$ can be computed by evaluating $j(\alpha)$ and computing the product,

$$H_D(x) = \prod_{\alpha}(x - j(\alpha)).$$

This can be done with complex floating point arithmetic, albeit to a high precision, as the coefficients of $H_D(x)$ must be integers. The product is taken over all α of the form

$$\alpha = (-b + \sqrt{-D})/(2a)$$

with $b^2 - 4ac = -D$, $|b| \leq a \leq \sqrt{|D|/3}$, $a \leq c$, $\gcd(a,b,c) = 1$ and if $|b| = a$ or $a = c$ then $b \geq 0$. In other words $ax^2 + bxy + cy^2$ is a primitive, reduced positive definite binary quadratic form of discriminant $-D$.

The need for high precision is crucial: as $\log|j(\alpha)| \approx \pi\sqrt{D}/a$, the coefficients of $H_D(x)$ can be huge. The precision needed will be around

$$10 + \binom{h_D}{\lfloor h_D/2 \rfloor}\frac{\pi\sqrt{D}}{\log 10}\sum_{\alpha}\frac{1}{a},$$

where the sum is over the same set of values of α as the above product.

The above is a very general outline of the global theory. To look at what happens over a finite field we need to localize all the above constructions. It will be assumed for simplicity that we are interested in curves defined over $\mathbb{F}_p$, where p is a large prime number.

VIII.2. Generating Curves over Large Prime Fields using CM

The method used is bound up with the arithmetic of complex quadratic fields, $K = \mathbb{Q}(\sqrt{-D})$. The number $-D$, which is a fundamental discriminant, is the basic input to the procedure. The method to be described is very fast if the class number, h_D, of K is small. However, some people have expressed concern at using a K with a small class number. The resulting curves may be more amenable to some future attack than more general K. On average it is expected that the class number of h_D will grow as $O(\sqrt{D})$, so small class numbers are in some sense 'special'. In addition, fields with small class numbers are well known to be easier to use in various algorithms (and hence possibly a future, as yet unknown, discrete logarithm algorithm).

We wish to construct a curve over $\mathbb{F}_p$ with complex multiplication by an order of discriminant $-D$. This means immediately that we are not going to construct supersingular elliptic curves. The j-invariant of such a curve is contained in $\mathbb{F}_p$ so a discriminant $-D$ is sought for which the Hilbert polynomial $H_D(x)$ has a root modulo p. So the field tower $\mathbb{Q} \subset K \subset H$ will collapse locally to either $\mathbb{Q}_p$ or a quadratic extension of $\mathbb{Q}_p$. In the latter case the prime p is inert in K, whilst in the former it splits (we shall ignore the case of ramified primes).

If the prime p is inert then there are no curves modulo p with complex multiplication by $\mathbb{Z}[\sqrt{-D}]$. Thus a prime which splits in p and for which the Hilbert class field is trivial when considered locally at p is needed. In other words we look for a principal ideal of K of norm p.

So given D we wish to know which prime numbers, p, split in K into prime ideals which are principal. Roughly $1/(2h_D)$ of the primes would be expected to have principal, degree one ideal divisors.

Such a prime will be one for which the diophantine equation

$$4p = x^2 + Dy^2$$

can be solved. This equation can be solved by the method of Cornacchia, which essentially computes the continued fraction expansion of the square root of a given rational. It is easy to see that solving $4p = x^2 + Dy^2$ is equivalent to solving $p = u^2 + dv^2$ and this task can be accomplished with the following method:

ALGORITHM VIII.1: **Cornacchia's Algorithm.**

INPUT: A square free integer d and a prime p.
OUTPUT: A solution to $p = u^2 + dv^2$, if one exists.

1. Let $p/2 < x_0 < p$ be a solution to $x^2 \equiv -d \pmod{p}$.
2. $p \leftarrow q_0 x_0 + x_1$, $k \leftarrow 0$.
3. Until $x_k^2 < p \leq x_{k-1}^2$ do:
4. $\quad x_k \leftarrow q_{k+1} x_{k+1} + x_{k+2}$, $k \leftarrow k+1$.
5. $u \leftarrow x_k$, $v \leftarrow \sqrt{(p - x_k^2)/d}$.
6. If $v \in \mathbb{Z}$ return (u, v), else return 'No Solution'.

To apply this we can repetitively try prime numbers until Cornacchia's algorithm finds a pair (x, y). By our previous comment, we expect to try $1/(2h_D)$ primes before a suitable one is found.

Given such a triple (x, y, p), compute

$$m = p + 1 \pm x.$$

These are going to be the possible group orders of the elliptic curves over $\mathbb{F}_p$ which we will try to construct. It can then be checked whether m is suitable, in that it has a large prime factor, is not equal to p and there is not a small number k such that $p^k \equiv 1 \pmod{m}$, the criteria set forth in Section V.7.

To explain why m is chosen in this manner, recall that the number of points on the curve is equal to

$$m = p + 1 - t$$

where t is the trace of Frobenius. Recall that $t = \alpha + \overline{\alpha}$, where α is an element of norm p in K. A solution to $x^2 + Dy^2 = 4p$ means that $\alpha = \pm(x + \sqrt{-D}y)/2$ is an element of norm p, with trace equal to $\pm x$. The orders $p + 1 \pm x$ will therefore be the orders of the elliptic curve and its quadratic twist.

For the field $\mathbb{F}_p$ and a group order m, it is required to build an elliptic curve over $\mathbb{F}_p$ with group order m. The main idea is contained in the following lemma.

LEMMA VIII.3. *The following hold for elliptic curves over* $\mathbb{F}_p$.

- *Every element in* $\mathbb{F}_p$ *is the* j*-invariant of an elliptic curve over* $\mathbb{F}_p$.
- *If* $D > 4$ *then all elliptic curves with given* j*-invariant,* $j \neq 0, 1728$*, over* $\mathbb{F}_p$ *are given by*

$$Y^2 = X^3 + 3kc^2X + 2kc^3$$

  where $k = j/(1728 - j)$ *and* c *is any element in* $\mathbb{F}_p$.
- *Suppose* E *and* E' *have the same* j*-invariant but are not isomorphic over the field* $\mathbb{F}_p$. *If* $j \neq 0$ *and* $j \neq 1728$*, then* E' *is the quadratic twist of* E *and if* $\#E = p + 1 - t$ *then* $\#E' = p + 1 + t$.
- *When* $j = 0$ *or* 1728*, additional cubic, quartic or sextic twists have to be considered. However, this case will be ignored, as in some sense such curves are special and hence should probably be avoided.*

Assume that $j \neq 0$ and $j \neq 1728$. In particular if E is given by

$$Y^2 = X^3 + aX + b$$

then E' can be given by

$$Y^2 = X^3 + ac^2X + bc^3$$

where c is any quadratic non-residue in $\mathbb{F}_p$. This means that if the j-invariant of a curve over $\mathbb{F}_p$ with order m can be constructed, then two candidate curves can be written down. Checking which one has the correct order is then done by means of randomly chosen curve points, as discussed in Section VI.2 (in fact, the problem here is slightly different, in that we have the order and need to distinguish between curves, as opposed to having to distinguish between candidate group orders for one curve; the method however is the same).

The problem has hence been reduced to computing which j-invariants can be the j-invariants of an elliptic curve over $\mathbb{F}_p$ with given number of points m and complex multiplication by the maximal order of $\mathbb{Q}(\sqrt{-D})$. As was seen above, such j-invariants must be the roots, modulo p, of the Hilbert polynomial $H_D(x)$.

All that remains is to compute $H_D(x)$, using the method given earlier. Finally, determining the roots of $H_D(x)$ over $\mathbb{F}_p$ can be accomplished using one of the many techniques for factoring polynomials over finite fields.

VIII.2.1. Examples. Two examples with small numbers are given, which can be used to test an implementation of the ideas involved.

Take $D = 7$ and look for a prime p such that

$$4p = x^2 + Dy^2$$

has a solution. Picking random primes, p, and applying Cornacchia's algorithm, we find a solution when $p = 7812216600826828873373526115 37$, which leads us to try to find an elliptic curve, over $\mathbb{F}_p$, with group order equal to

$$m = 781221660082681210712714541668,$$

which is four times an odd prime. The class number of $K = \mathbb{Q}(\sqrt{-D})$ is one, so the Hilbert polynomial $H_7(x)$ has degree one. It is found to be equal to

$$x + 3375$$

which clearly has a root modulo p. An elliptic curve with j-invariant,

$$j_E = -3375 \equiv 781221660082682887337352608162$$

is needed. Up to $\mathbb{F}_p$-isomorphism, there are two such curves, given by

$$\begin{aligned} E : Y^2 = X^3 &+ 384410658135923325515205253294X \\ &+ 777088212145737475235038576554 \end{aligned}$$

and

$$\begin{aligned} E' : Y^2 = X^3 &+ 586337137088968521507562977329X \\ &+ 470612877688284093511930750213. \end{aligned}$$

It is then a simple matter, which is left as an exercise, to determine which of the above two curves has the group order m. The whole computation takes a fraction of a second. It is in fact the latter of the above two elliptic curves which has order m.

As a second example take $D = 292$, which has class number four. For the prime $p = 4710640177146485817437161152 53$ the equation $x^2 + Dy^2 = 4p$ can be solved and from the solutions it is deduced that there are elliptic curves of group order equal to

$$m = 471064017714647630725498582802.$$

The Hilbert polynomial $H_D(x)$ is given by

$$\begin{aligned} x^4 &- 206287709860428304608000x^3 \\ &- 9369362251192903875949706611 2000000x^2 \\ &+ 4552155138637938536962996838 4000000000x \\ &- 3802594610425124047799906426 88000000000000 \end{aligned}$$

which has four roots modulo p. The four roots are the values of four j-invariants of elliptic curves over $\mathbb{F}_p$ with group orders equal to m. One such root is $j = 9529816310558554289907682343 5$, from which the following two elliptic curves are computed:

$$\begin{aligned} E : Y^2 = X^3 &+ 4692684364282467257810351342 77X \\ &+ 1558242850472816232727847177 67 \end{aligned}$$

and

$$\begin{aligned} E' : Y^2 = X^3 &+ 3546187395733478131233890933 24X \\ &+ 3142517785930543625908799545 74. \end{aligned}$$

The second curve has order m, while the first has order $2(p+1) - m$. The other three roots of $H_D(x) \pmod p$ also give rise to elliptic curves over $\mathbb{F}_p$ with order m. These are

$$\begin{aligned} E'' : Y^2 = X^3 &+ 2260375678353386115691921988 97X \\ &+ 1506917118902257410461281325 98, \\ E''' : Y^2 = X^3 &+ 4705690056267710307215285582 11X \\ &+ 7331063557219604895221098683, \\ E'''' : Y^2 = X^3 &+ 3063530651060261108031053080 74X \\ &+ 2042353767373507405354035387 16. \end{aligned}$$

In the Appendix we give a few more examples over larger finite fields.

VIII.3. Weber Polynomials

There are two problems with the above method. The first is that the curves produced are in some sense 'special'. They will have relatively small class numbers and so could be more amenable to some future unknown attack. As a general principle, it is believed that choosing random curves, and computing their group order via Schoof's algorithm, is more likely to produce curves which are resistant to specialized attacks. As of the writing of this book, there is virtually no evidence to support this belief, but it enjoys wide support in the community.

The second problem with the above method is that to find the j-invariants the Hilbert polynomials, $H_D(x)$, have to be computed. It has been noted that this requires computing the coefficients to what can be a prohibitively large precision. One way to get around this is to compute the minimal polynomial of another generator of the Hilbert class field which has a known algebraic relationship to $j(\tau)$. The advantage is that this second polynomial may have much smaller coefficients which would allow us to use a much smaller precision. In this section a possible solution to this second problem is considered.

Define the following Weber functions, using Dedekind's η-function, $\eta(z)$:

$$h(\tau) = \zeta_{48}^{-1}\frac{\eta((\tau+1)/2)}{\eta(\tau)},\ h_1(\tau) = \frac{\eta(\tau/2)}{\eta(\tau)},\ h_2(\tau) = \sqrt{2}\frac{\eta(2\tau)}{\eta(\tau)},$$

$$\gamma_2(\tau) = \frac{h(\tau)^{24}-16}{h(\tau)^8},\ \gamma_3(\tau) = \frac{(h(\tau)^{24}+8)(h_1(\tau)^8 - h_2(\tau)^8)}{h(\tau)^8},$$

where $\zeta_n = e^{2\pi i/n}$. These functions are not all algebraically independent since they are all related to j via the equations (see [7])

$$j = \frac{(h^{24}-16)^3}{h^{24}} = \frac{(h_1^{24}+16)^3}{h_1^{24}} = \frac{(h_2^{24}+16)^3}{h_2^{24}} = \gamma_2^3 = \gamma_3^2 + 1728.$$

Weber calls $\mu(\tau)$ a class invariant if $\mu(\tau)$ lies in the Hilbert class field of $\mathbb{Q}(\tau)$. Clearly $j(\tau)$ is a class invariant. However, using the Weber functions we can determine a lot more class invariants. These give rise to polynomials, usually denoted by $W_D(x)$, using almost the same method as we used to compute $H_D(x)$. Finding the roots of these new polynomials, which will hopefully have smaller coefficients, will then allow us to recover the j-invariant.

Atkin and Morain [7] suggest using the following choices of class invariants to produce the Weber polynomials: Remember $-D$ is a fundamental discriminant and we have that d is the square free positive integer such that $\mathbb{Q}(\sqrt{-D}) = \mathbb{Q}(\sqrt{-d})$. The following conditions are applied in turn (in other words the condition on D being divisible by 3 takes priority).

- If $D \equiv 3 \pmod 6$ use $\mu = \sqrt{-D}\gamma_3(\tau)$.
- If $D \equiv 7 \pmod 8$ use $\mu = h(\tau)/\sqrt{2}$.
- If $D \equiv 3 \pmod 8$ use $\mu = h(\tau)$.
- If $d \equiv \pm 2 \pmod 8$ use $\mu = h_1(\tau)/\sqrt{2}$.
- If $d \equiv 5 \pmod 8$ use $\mu = h(\tau)^4$.
- If $d \equiv 1 \pmod 8$ use $\mu = h(\tau)^2/\sqrt{2}$.

The only problem here is the case when $D \equiv 3 \pmod 8$ and $D \not\equiv 3 \pmod 6$. In that case the degree of $W_D(x)$ is $3h_D$ and not h_D. This could be a problem, so one could just ignore such discriminants. Atkin and Morain also give detailed descriptions of how to compute the various Weber polynomials and other alternative class invariants to use.

As an example of the advantage that using Weber polynomials can bring, Atkin and Morain give the following example:

$$H_{23}(x) = x^3 + 3491750x^2 - 5151296875x + 23375^3$$

while

$$W_{23}(x) = x^3 - x - 1.$$

VIII.4. Further Discussion

Given a fundamental discriminant, $-D$, it has been shown how to choose p, obtain a desirable group order and then construct an elliptic curve with that group order. It was also proposed that one should randomly choose prime numbers, p, and then solve the relevant diophantine equation using Cornacchia's method. It is often far better to choose the prime first and then the discriminant, since some primes are more amenable to implement the arithmetic needed for an elliptic curve system.

Some improvement can be obtained by examining the possible splitting behaviour of the rational prime p in the quadratic extension $K = \mathbb{Q}(\sqrt{-d})$ and its Hilbert class field. This reduces to the following result:

LEMMA VIII.4. *Let d be square free and p a prime number such that we can find a solution to the equation*

$$p = x^2 + dy^2,$$

then we have the following.

- *If* $p \equiv 3 \pmod 8$ *then* $D \equiv 2, 3$ *or* $7 \pmod 8$.
- *If* $p \equiv 5 \pmod 8$ *then* $D \equiv 1 \pmod 2$.
- *If* $p \equiv 7 \pmod 8$ *then* $D \equiv 3, 6$ *or* $7 \pmod 8$.

In particular we must have $(\frac{-d}{p}) = (\frac{-D}{p}) = 1$.

In the case of characteristic two the method is similar. First, assuming the finite field has cardinality $q = 2^n$, solve the equation

$$2^{n+2} = x^2 + Dy^2$$

for some positive square free number D. The arithmetic of the quadratic field $\mathbb{Q}(\sqrt{-D})$, will again be used. However, the class number h_D must now be divisible by n. That way, when we localize the field $\mathbb{Q}(j_D)$ which is of degree h_D and Galois, there is the chance of obtaining a local extension of $\mathbb{Q}_2$ of degree n. Only in this way will the residue field be of size 2^n.

Again we have an integer D, a proposed group order m and a field of definition $\mathbb{F}_{2^n}$. If the Hilbert polynomial modulo 2 has a degree n irreducible factor then this polynomial can be used to define the extension $\mathbb{F}_{2^n}$ over $\mathbb{F}_2$. If the Hilbert polynomial has no such factor modulo 2 then another D and n are tried.

Let the degree n irreducible factor of H_D, over $\mathbb{F}_2$, be denoted by $p(x)$. Let α denote a root of $p(x)$ in $\mathbb{F}_{2^n}$. This is then the j-invariant of an elliptic curve over $\mathbb{F}_{2^n}$ which has CM by an order of $\mathbb{Q}(\sqrt{-D})$. We then only need to generate a curve with given j-invariant and actually test whether it has the correct order. As mentioned earlier, more details on this can be found in [**73**].

CHAPTER IX

Other Applications of Elliptic Curves

In this chapter we discuss a number of additional applications of elliptic curves in cryptography, namely factoring, primality proving and proving the equivalence of the Diffie–Hellman problem to the DLP. Only the central ideas of each application are discussed. More comprehensive descriptions can be found in the references cited.

IX.1. Factoring Using Elliptic Curves

We shall give a brief description of Lenstra's [78] *elliptic curve factoring method*, usually referred to as ECM. Let N be a number which is to be factored and let p denote some, as yet unknown, prime factor of N.

To introduce the elliptic curve method, consider first Pollard's $p-1$ method of factoring. For convenience, assume that N is of the form $N = p \cdot q$, the product of two prime numbers. The group $(\mathbb{Z}/N\mathbb{Z})^*$ decomposes (via the CRT) as

$$(\mathbb{Z}/N\mathbb{Z})^* \cong \mathbb{F}_p^* \times \mathbb{F}_q^*.$$

Take an element $a \in (\mathbb{Z}/N\mathbb{Z})^*$, and raise it to the power of a multiple of $p-1$, say $\lambda(p-1)$, then

$$a^{\lambda(p-1)} \equiv 1 \pmod{p}.$$

It can be expected that

$$\gcd(a^{\lambda(p-1)} - 1, N) = p.$$

The problem with this is that we need to know $p-1$ to recover p. However, if we make the assumption that $p-1$ is 'smooth', i.e. all of its divisors are less than some given bound, then, if a large smooth number M is chosen, there is a chance that $p-1$ will divide M. In such a situation p can be recovered by computing

$$\gcd(a^M - 1, N) = p.$$

For example, suppose $N = 12628003$ and we choose $a = 2$ and $M = 20!$ Then $a^M \pmod{N} = 2804399$ and

$$\gcd(a^M - 1, N) = \gcd(2804398, 12628003) = 2053.$$

We deduce the factorization $N = 2053 \times 6151$. This example works since, on setting $p = 2053$, we see that $p - 1 = 2^2 \cdot 3^3 \cdot 19$ divides $20!$ and $q - 1$ does not divide $20!$

In practice one can make various improvements to this strategy (see [**29**]). For example we can choose M to be the number

$$M(k) = \text{lcm}(1, 2, \ldots, k).$$

However, it is explicitly relying on the fact that $\mathbb{F}_p^*$ has a smooth group order, but not all prime factors of a large number satisfy this requirement. For example if $N = 4268347$ and we choose $a = 2$ and $M = M(20)$ then we do not obtain a factor of N. This is because N is of the form $p \cdot q$, for primes p and q, and $p - 1$ and $q - 1$ in this example both have prime factors larger than 19.

This is where elliptic curves are effective. We first note that an elliptic curve over the ring $\mathbb{Z}/N\mathbb{Z}$ is a curve with coefficients in $\mathbb{Z}/N\mathbb{Z}$ and whose coordinates also lie in $\mathbb{Z}/N\mathbb{Z}$. We can define a natural group law on such curves, although one usually gives these in terms of projective coordinates so as to cope with the occurrence of zero divisors in $\mathbb{Z}/N\mathbb{Z}$. An elliptic curve over $\mathbb{Z}/N\mathbb{Z}$ splits via the CRT into

$$E(\mathbb{Z}/N\mathbb{Z}) \cong E(\mathbb{F}_p) \times E(\mathbb{F}_q).$$

For every value of N we have various choices of E and a chance of finding a curve for which $\#E(\mathbb{F}_p)$ is smooth for some prime factor, p, of N.

First we find an elliptic curve with a projective point on it modulo N:

$$\mathcal{C}_{a,b} : Y^2 Z = X^3 + aXZ^2 + bZ^3$$

and $(x, y, z) \in \mathcal{C}_{a,b}(\mathbb{Z}/N\mathbb{Z})$. We assume N is coprime to Δ and so, as p does not divide Δ, we know that $\mathcal{C}_{a,b}$ has good reduction at p. By Hasse's Theorem, considering $\mathcal{C}_{a,b}$ as a curve over $\mathbb{F}_p$, it has order N_p, where

$$|N_p - (p + 1)| < 2\sqrt{p}.$$

A number k is chosen and

$$(x_k, y_k, z_k) \equiv [M(k)](x, y, z) \pmod{N}$$

computed. If N_p divides $M(k)$ then p divides z_k and a factor of N might be found by taking $\gcd(z_k, N)$. In practice this factor will be found whilst computing $[M(k)](x, y, z)$ as some inversion will become impossible due to the presence of zero divisors in the ring $\mathbb{Z}/N\mathbb{Z}$.

The method relies on the fact that the smooth number $M(k)$ is a multiple of the group order $E(\mathbb{F}_p)$. There is considerable freedom here; we can choose the coefficients a and b from a large number of possibilities. However, for prime numbers of between 20 and 40 decimal digits one is likely to write down an elliptic curve with a smooth group order after not too long a time (see [**78**] and [**89**]).

To see why this works efficiently, a few facts about elliptic curves modulo p are noted.

LEMMA IX.1. *There is a positive constant* α *such that* αp^2 *of all pairs* $(a, b) \in \mathbb{F}_p \times \mathbb{F}_p$, *with* $4a^3 + 27b^2 \neq 0$, *give a curve with*

$$\#C_{a,b}(\mathbb{F}_p) \in (p - \sqrt{p}, p + \sqrt{p}),$$

and such group orders in this range are distributed in an approximately uniform manner.

With every new elliptic curve used, we have a probability of about α of choosing a curve with order in $(p - \sqrt{p}, p + \sqrt{p})$. Choosing the curve can be interpreted as choosing a random integer T from a uniform distribution on $(p - \sqrt{p}, p + \sqrt{p})$. The elliptic curve method will have a very good chance of finding a factor of our number, with this choice of T, if T has a good chance of dividing $M(k)$.

The method is more likely to work for larger values of k, and hence larger values of $M(k)$. However, the larger the value of k the more work is needed for each curve. A good strategy is to start with a medium size value of k, at least one for which $M(k) > p$. Then, if a factor is not found after a few attempts, the value of k is increased and the procedure is repeated until it is successful. Using this idea of increasing the value of k, the complexity of the method is given by

$$O(L_p(0.5, \sqrt{2})),$$

where we recall from Chapter I that the function L_p is given by

$$L_p(v, c) = \exp(c(\log p)^v (\log \log p)^{(1-v)}).$$

For numbers of the form $N = pq$, with p and q of order $\sqrt{N}$ and $N > 10^{80}$, the elliptic curve method is inefficient compared to the quadratic or number field sieve methods, even though, for integers of this form, ECM has approximately the same asymptotic complexity as the quadratic sieve method. This is because the basic operations in ECM are far more complicated than those used in the quadratic sieve.

What makes the ECM the most successful factoring algorithm known on hundred digit numbers is that it is very rare for a random hundred digit number to be of the form pq, where p and q are of roughly the same size. So for a random integer of around one hundred digits ECM should find the prime factors before a more advanced method such as the quadratic or number field sieves. However, in cryptography one is usually interested in the types of numbers for which p and q do have roughly the same size. Hence it could appear that the uses of ECM in cryptography are very limited.

This is not true; the large prime variations of both the quadratic and number field sieve algorithms require the factorization of numbers as a subprocedure. It is at this stage that the ECM method can be applied with some success. In addition it will be seen later that primality proving algorithms also require the factorization of auxiliary numbers. Since one could expect that any auxiliary number produced by an algorithm should be of the form of a random number of the required size, one would expect that ECM would be able to factor such numbers, and indeed it usually does.

IX.2. The Pocklington–Lehmer Primality Test

As mentioned previously, in Chapter VIII, the CM method was first used to construct elliptic curves in the context of a primality proving algorithm. This elliptic curve primality proving (ECPP) method is itself based on the Pocklington–Lehmer primality test ([**123**],[**132**]). To introduce ECPP, the Pocklington–Lehmer $N-1$ primality test is therefore first discussed. It is then shown how, by replacing the group $\mathbb{F}_p^*$ by the group $E(\mathbb{F}_p)$, a more powerful primality test can be obtained. Only the very basic design of the primality test is considered, leaving the reader to consult the literature for further optimizations, improvements and enhancements.

Assume that the number N has already passed the Miller–Rabin pseudoprime test ([**102**], [**130**]). There is then some confidence that the number N really is a prime. We merely want to verify this confidence. However, we will achieve more than this. An output from the primality proving program will be produced which will convince someone else that the number is prime, without them having to run the algorithm again. In other words the algorithm should produce a 'proof' or certificate of the primality of N: the primality of the number is easily verified, in an irrefutable manner, with the additional information provided.

Consider the following theorem.

THEOREM IX.2. *Suppose N is an integer and p a prime divisor of $N-1$, with p^e being the largest power of p that divides $N-1$. Also suppose that there is an a such that*

$$a^{N-1} \equiv 1 \pmod{N}$$

and

$$\gcd(a^{(N-1)/p}, N) = 1.$$

Then if q is any divisor of N we have

$$q \equiv 1 \pmod{p^e}.$$

This theorem can be turned into a primality test by using the following corollary:

COROLLARY IX.3. *Write $N-1$ as AB where A and B are coprime, the factorization of A is completely known and $A > \sqrt{N}$. For each prime factor, p, of A we can find an a_p such that*

$$a_p^{N-1} \equiv 1 \pmod{N} \text{ and } \gcd(a_p^{(N-1)/p} - 1, N) = 1,$$

if and only if N is prime.

To prove the primality of N we need to partially factor $N-1$ into A and B of Corollary IX.3 and then for each prime factor of A find an integer, a_p, which satisfies the conditions above. It does not matter how such values of a_p are found. Once found, their existence will guarantee the primality of N.

One problem that may be encountered is that in factoring $N-1$ the primality of another number will have to be established, and so on. This gives rise to the so called **down run** process whereby the proof of primality of one number is dependent on the proof of primality of another and so on.

As an example the primality of $N = 105554676553297$ will be proven. The integer $N-1$ has the following factors:

$$N - 1 = 2^4 \times 3 \times 1048583 \times 2097169.$$

If we set $A = 3 \times 1048583 \times 2097169$ and $B = 2^4$ then $A > \sqrt{N}$ and $\gcd(A, B) = 1$. Notice that if we set

$$a_3 = a_{1048583} = a_{2097169} = 2$$

then the primality of N can be established using the above corollary, assuming $p = 1048583$ and $q = 2097169$ are themselves primes.

This leads us to perform the following down run. Write

$$p - 1 = 2 \times 29 \times 101 \times 179.$$

On setting $A = 29 \times 101$ and $B = 358$, we notice that $a_{29} = a_{101} = 2$ will prove the primality of p. Assume here that the primality of numbers less than 1000 is proved by table lookup. We then need to prove the primality of q:

$$q - 1 = 2^4 \times 3 \times 43691.$$

It is seen that taking $a_3 = 5$ and $a_{43691} = 2$ will prove the primality of q. We also need to prove the primality of 43691, which is done in a similar way.

The following certificate of primality is thus obtained:

```
105554676553297
3        2
1048583  2
2097169  2
----------
1048583
29       2
101      2
----------
2097169
3        5
43691    2
----------
43691
257      3
----------
```

The main problem with this method is that the partial factorization of $N-1$ is needed. Suppose it is wished to prove the primality of a modulus to be used in an RSA scheme, or to define an elliptic curve scheme over $\mathbb{F}_p$. This means that p could be of the order of 10^{100}. Hence it may be required to factor a number of around one hundred decimal digits, and this can be a non-trivial task without a large amount of computing power available.

IX.3. The ECPP Algorithm

In the last section the group $(\mathbb{Z}/N\mathbb{Z})^*$ was used to prove the primality of N. Since N is believed to be prime, the order of $(\mathbb{Z}/N\mathbb{Z})^*$ is expected to be $N-1$. The method will work if the number $N-1$ is suitably smooth, but we could be working with a group order which is not smooth. In this situation using another group which has a chance of having a smooth group order will improve the situation.

Just as with the elliptic curve factoring method, the group $(\mathbb{Z}/N\mathbb{Z})^*$ can be replaced with the group of points on an elliptic curve over $\mathbb{Z}/N\mathbb{Z}$. This last group has a chance of having a smooth order, and even if it does not curves can continue to be chosen until one is found with a smooth order. This is the idea behind the elliptic curve primality proving algorithm, which is now discussed.

The method is due to Goldwasser and Kilian [**46**] who gave the following analogue of the method of Pocklington and Lehmer.

THEOREM IX.4. *Suppose N is an integer coprime to six and larger than one. Let E denote an elliptic curve over $\mathbb{Z}/N\mathbb{Z}$. Assume that one can compute an integer m which has a prime divisor q with*

$$q > (N^{1/4}+1)^2.$$

If a point $P \in E(\mathbb{Z}/N\mathbb{Z})$ can be found such that

$$[m]P = \mathcal{O} \text{ and } [m/q]P \neq \mathcal{O},$$

then N is prime. Note if neither of the above multiplications is possible then a non-trivial factor of N has been found, just as with the ECM factoring method, and so N is not prime.

The following result implies that, once a suitable order has been found, such a point P must exist.

LEMMA IX.5. *Let E denote an elliptic curve over $\mathbb{Z}/N\mathbb{Z}$ with order equal to m and with N prime. If m has a prime divisor q such that*

$$q > (N^{1/4}+1)^2,$$

then there exists a point $P \in E(\mathbb{Z}/N\mathbb{Z})$ such that $[m/q]P \neq \mathcal{O}$.

All that remains is to keep producing random elliptic curves and calculating their group orders until one is found which will prove the given number,

N, is prime. If a suitable point P on the elliptic curve is found, the result follows. Just as with the Pocklington–Lehmer method a down run strategy can be adopted if in such a process it is needed to prove another number is prime.

Goldwasser and Kilian suggested that as it is very likely that N is prime, Schoof's algorithm can be used to compute the order of $E(\mathbb{Z}/N\mathbb{Z})$. This order can then be trial divided to see if it divisible by a large prime q and one can proceed from there. As Schoof's algorithm runs in polynomial time this gives us a probabilistic polynomial time primality test. Clearly the output from the algorithm is a certificate of the primality of N which can be checked in a shorter time than it took to generate it.

Atkin and Morain [**7**] noticed that it would be more efficient to use the CM-theory of elliptic curves to generate the required curve. However instead of determining the group order of a random curve they find a curve given a group order.

They first find a discriminant D for which there exist elliptic curves over $\mathbb{F}_N$ (assuming N is prime) with complex multiplication by an order of $\mathbb{Q}(\sqrt{-D})$. They can then compute the possible group orders of curves with this complex multiplication structure. It is determined which (if any) of these group orders possess a large prime factor, q, of the form above. This may require calling the primality proving program recursively in a down run strategy. Finally, using the CM-theory mentioned in Chapter VIII, a curve with the required order can be found, and a point can be constructed on the curve with the required properties.

This method has been implemented and is very successful. It can prove the primality of numbers with over one thousand decimal digits. However, it is still not a deterministic polynomial time primality test, although it is very practical. A deterministic polynomial time algorithm has been given by Adleman and Huang [**3**]. This latter test has to our knowledge never been implemented, perhaps because it uses the arithmetic of Jacobians of hyperelliptic curves of genus two. From a practical view point there seems no point in preferring the method of Adleman and Huang to ECPP.

IX.3.1. Example. Consider proving the primality of $p = 2^{100} + 277$. Using the methods developed to find curves using the CM method the elliptic curve

$$\begin{aligned} Y^2 = X^3 \; &+ \; 169317673849406496638751929789X \\ &+ \; 535428649309014131591402355077 \end{aligned}$$

over $\mathbb{F}_p$, is found, which has order $m = 1267650600228230776357544186344$. After trial division it is seen that m has an 81-bit cofactor, which is probably prime. Call this cofactor $p_1 = 1764763222984205716119937$. Assuming that p_1 is prime the result of Goldwasser and Kilian will show that p is prime on noticing that the point

$$P = (12231165171072343718908796 08558,\ 34881870097669254769721966 5601)$$

satisfies $[m]P = \mathcal{O}$ but $[m/q]P \neq \mathcal{O}$.

A down run strategy is started to prove the primality of p_1 and so on. It is left to the reader to verify that the following is a valid certificate (with an obvious notation):

```
1267650600228229401496703205653
1693176738494064966387519297 89   5354286493090141315914023550 77
1223116517107234371890879608558   3488187009766925476972196656 01
1267650600228230776357544186344
1764763222984205716119937
-------------------------
1764763222984205716119937
1237106009019141934754397 8247373393460946231695 98
4985662653836856558503 76   1698160958763013389415626
1764763222981587729747968
21321838780409719
-------------------------
21321838780409719
5979072666605065   11093328037873283
12289991207526417  5086330291908954
21321839059327264
636820759
-------------------------
636820759
572504044 593942949
442683250 159049258
636870910
37397
```

This last number is easily seen to be prime as it is not divisible by any prime less than 193.

IX.4. Equivalence between DLP and DHP

In this section we describe the ideas of Maurer, Wolf and Boneh ([**94**], [**95**], [**18**]) for showing the equivalence between the DLP and the DHP for various special classes of groups.

To recap, given a finite abelian group G, the DLP is: given $g, h \in G$, find the integer m, if it exists, such that

$$h = g^m.$$

We write $m = \mathrm{DL}_g(h)$. The DHP is: given $g^a, g^b, g \in G$, determine the element

$$g^{ab} = \mathrm{DHP}_g(g^a, g^b).$$

Clearly if we can solve the DLP then we can solve the Diffie–Hellman problem, since we can evaluate the function DHP_g by a polynomial number of calls to the function DL_g and a polynomial number of group operations in G. The hard part is showing that we can solve the DLP using a polynomial number of group operations and a polynomial number of calls to the function, DHP_g, which solves the DHP.

We outline the various ideas of the equivalence and leave the reader to consult the relevant papers for details. We assume that G is cyclic of prime order and is generated by g. Let the order of g be p.

First we construct an elliptic curve over the field $\mathbb{F}_p$ of the form

$$E : Y^2 = X^3 + AX + B$$

which is cyclic and whose order is $P(\log p)$-smooth, where $P(\log p)$ is some polynomial in $\log p$. In particular

$$\#E(\mathbb{F}_p) = \prod_{q_j \leq P(\log p)} q_j^{f_j}.$$

With current knowledge it is unclear how this is actually done or indeed whether such curves always exist. It is this last fact which makes the following method only apply to cyclic groups with certain prime orders.

We are given $h \in G = \langle g \rangle$ and we wish to compute $m \in \mathbb{Z}$ such that $h = g^m$. We let $P = (u, v)$ denote a generator of the cyclic elliptic curve $E(\mathbb{F}_p)$. Let $m' = m \pmod{p} \in \mathbb{F}_p$ and think of m' as the x-coordinate of some, unknown, elliptic curve point $Q = (m', n')$. Such a value of n' may not exist, in which case we need to take the quadratic twist of E. However, for the purposes of explanation let us assume that such a value of n' could be found (if we knew m'). Now suppose we can solve the following ECDLP on $E(\mathbb{F}_p)$:

$$Q = (m', n') = [s]P.$$

After we have determined s we can compute $[s]P$ and hence determine m' and the original DLP would have been solved. The main obstacle to this approach is determining s when we do not even know Q.

The trick is to use the fact that we are allowed to make a polynomial number of calls to the function DHP_g above. Firstly, given g^m we can compute

$$g^{m^3+Am+B} = g^{m(m^2+A)+B} = g^z$$

using two calls to DHP_g and a polynomial number of group operations. We can test whether z is a quadratic residue modulo p using the equivalence

$$z^{(p-1)/2} \equiv 1 \pmod{p} \Leftrightarrow g^{z^{(p-1)/2}} = g.$$

This can be tested using a polynomial number of group operations and calls to DHP_g. Hence we can determine whether we need to take a quadratic twist of the curve without actually knowing what the value of m' is. Again suppose we do not need to take a quadratic twist (for the case where we do you should consult the above mentioned papers).

Using the function DHP_g we can then compute g^n where

$$g^{n^2} = g^z = g^{m^3+Am+B},$$

using a technique similar to the method of Tonelli and Shanks in Chapter II. So although we do not know (m', n') we do know

$$(g^{m'}, g^{n'}).$$

Suppose $(a,b), (c,d) \in E(\mathbb{F}_p)$ but we only know

$$(g^a, g^b), (g^c, g^d) \in G \times G;$$

then we can compute, using the function DHP_g, the group operations in G and the formulae for the group law of the elliptic curve in terms of two group elements

$$(g^e, g^f) \in G \times G$$

such that on the curve $E(\mathbb{F}_p)$ we have

$$(e,f) = (a,b) + (c,d).$$

This is done using the standard elliptic curve group law formulae but replacing multiplications modulo p by calls to the function DHP_g.

Since the group order of $E(\mathbb{F}_p)$ is $P(\log p)$-smooth we can easily solve the ECDLP

$$Q = (m', n') = [s]P$$

using Pohlig–Hellman, the BSGS algorithm and the 'virtual' group operation defined above. We can do this using a polynomial (in $\log p$) number of group operations in G and a polynomial number of calls to the function DHP_g.

Hence the Diffie–Hellman problem and the DLP are equivalent if we can find an elliptic curve over $\mathbb{F}_p$ with the required smooth number of points. Using these ideas one can prove:

THEOREM IX.6 (Maurer and Wolf, [**95**]). *Let G be a cyclic group of prime order, p. Let B denote a smoothness bound which is polynomial in* $\log p$. *The Diffie–Hellman problem in G and the DLP in G are polynomial time equivalent if one of the following expressions is B-smooth:*

$$p \pm 1,$$

$$p+1 \pm 2a\ ,\ p+1 \pm 2b$$

where $p \equiv 1 \pmod 4$, $p = a^2 + b^2$ *and* $a + b\sqrt{-1} \equiv 1 \pmod{2 + 2\sqrt{-1}}$;

$$p+1 \pm 2a\ ,\ p+1 \mp a \pm 2b\ ,\ p+1 \pm (a+b)$$

where $p \equiv 1 \pmod{p}$, $p = a^2 - ab + b^2$ *and* $a + b\omega \equiv 2 \pmod{3}$ *with* $\omega^2 + \omega + 1 = 0$.

CHAPTER X

Hyperelliptic Cryptosystems

In this final chapter the generalization of systems based on elliptic curves to systems based on hyperelliptic curves is considered. The cryptography is the same: the only change is the replacement of the group of points on an elliptic curve by the group of points of the Jacobian of a hyperelliptic curve. Hyperelliptic curves were first proposed for use in cryptography by Koblitz [**63**].

X.1. Arithmetic of Hyperelliptic Curves

Let C denote a hyperelliptic curve of genus g defined over $\mathbb{F}_q$, with imaginary quadratic function field K. A hyperelliptic curve, C, of genus g can be given in the form

$$C : Y^2 + H(X)Y = F(X)$$

where $F(X)$ is a monic polynomial of degree $2g+1$ and $H(X)$ is a polynomial of degree at most g. Both $H(X)$ and $F(X)$ have coefficients in $\mathbb{F}_q$. Such a curve is non-singular if for no point on $C(\overline{\mathbb{F}_q})$ does there exist a point for which the two partial derivatives,

$$2Y + H(X) \text{ and } H'(X)Y - F'(X),$$

simultaneously vanish. It will always be assumed that the curve C is non-singular. In odd characteristic fields it will always be assumed that $H(X) = 0$ and that $F(X)$ is square free.

A divisor on a curve is a formal sum of points

$$D = \sum_{P \in C(\overline{\mathbb{F}}_q)} n_P P$$

where $n_P \in \mathbb{Z}$ and all but finitely many of the n_P are zero, the degree of a divisor is defined to be $\sum n_P$. A divisor is called effective if $n_P \geq 0$ for all P and is called rational if it is stable under the action of the absolute Galois group over $\mathbb{F}_q$.

Every function on the curve gives rise to a divisor of degree zero, consisting of the formal sum of the poles and zeros of the function. Such divisors are called principal. The group of rational divisors of degree zero modulo principal divisors forms the *Jacobian* of C over $\mathbb{F}_q$, denoted by $J_C(\mathbb{F}_q)$. This is a finite abelian group and forms the basis of the cryptographic schemes based on hyperelliptic curves.

A divisor on C will be called semi-reduced if it is effective and if, when a point P occurs in the support of the divisor, then the point $\tilde{P}$ does not, where $\tilde{P}$ denotes the image of P under the hyperelliptic involution. A semi-reduced divisor, which is defined over $\mathbb{F}_q$, can be represented by two polynomials $a, b \in F_q[x]$ which satisfy

(i) $\deg b < \deg a$,
(ii) b is a solution of the congruence $b^2 + Hb \equiv F \pmod{a}$.

Such a divisor will be denoted by $\operatorname{div}(a, b)$, and it represents the $\mathbb{F}_q$-rational divisor

$$\sum_{x_i} m_i(x_i, b(x_i))$$

where the sum is over all roots x_i of a, each root having multiplicity m_i.

The Jacobian, J_C, can be represented uniquely by reduced divisors. A reduced divisor is a semi-reduced divisor as above but of degree less than or equal to g. Hence the polynomial a above will have degree less than or equal to g. The identity of the group law on J_C is given by $\mathcal{O} = \operatorname{div}(1, 0)$, and addition can be performed using the well known algorithm of Cantor and Koblitz (see [**24**] and [**63**]). Cantor's algorithm is equivalent to the usual combination and reduction algorithm of binary quadratic forms. In the function fields under consideration a divisor is essentially equivalent to a binary quadratic form, a fact we will return to later.

ALGORITHM X.1: **Cantor's Algorithm.**

INPUT: Two reduced divisors $D_1 = \operatorname{div}(a_1, b_1)$ and $D_2 = \operatorname{div}(a_2, b_2)$.
OUTPUT: The reduced divisor $\operatorname{div}(a_3, b_3) = D_1 + D_2$.

1. Perform two extended gcd computations to compute
   $d = \gcd(a_1, a_2, b_1 + b_2 + H) = s_1 a_1 + s_2 a_2 + s_3(b_1 + b_2 + H)$.
2. $a_3 \leftarrow a_1 a_2 / d^2$,
3. $b_3 \leftarrow (s_1 a_1 b_2 + s_2 a_2 b_1 + s_3(b_1 b_2 + F))/d \pmod{a_3}$.
4. While $\deg a_3$ is greater than the genus of C do:
5. $\quad a_3 \leftarrow (F - H b_3 - b_3^2)/a_3$,
6. $\quad b_3 \leftarrow -H - b_3 \pmod{a_3}$.
7. Return $\operatorname{div}(a_3, b_3)$.

It is easy to see that the degree of a_3 will monotonically decrease as we process this while loop and so eventually a reduced divisor will be obtained. The initial steps are analogous to the composition of binary quadratic forms, while the final while loop is analogous to the reduction method for binary quadratic forms. For an analysis of the complexity of the above method, improvements and an extension to real quadratic function fields see [**118**] and [**119**].

For the rest of this section it will be assumed, for simplicity, that $\mathbb{F}_q$ has odd characteristic and that $H(X) = 0$. As K is a quadratic function field, prime divisors, P, in K come in one of three varieties. Let p denote the prime of $\mathbb{F}_q[x]$ which lies below P, in which case we have:

- **P ramifies**
  In this case p divides F and there is only one ramified prime divisor, P, lying above p. Denote this prime divisor by $\mathrm{div}(p, 0)$.
- **P is inert**
  In this case p does not divide F and there is no solution to the equation
  $$y^2 \equiv F(x) \pmod{p}$$
  in the field $L = \mathbb{F}_q[x]/(p)$. Whether such a solution exists can be determined by either using a standard generalization of the Legendre symbol or factoring $y^2 - F$ over the field L.
- **P splits**
  As in the inert case p does not divide F but now the equation
  $$y^2 \equiv F(x) \pmod{p}$$
  has two solutions, r_1 and r_2, both of degree less than p. The prime, p, then splits into the two divisors
  $$P = \mathrm{div}(p, r_1) \text{ and } \tilde{P} = \mathrm{div}(p, r_2).$$
  The values of the polynomial r_1 (and hence r_2) can be determined either by factoring $y^2 - F$ over the field $L = \mathbb{F}_q[x]/(p)$, or by using an obvious generalization of the algorithm of Tonelli and Shanks (see Algorithm II.8).

X.2. Generating Suitable Curves

Just as in the case of elliptic curves, there are many ways one could theoretically proceed if one wanted to produce curves suitable for use in cryptography. The order $\#J_C(\mathbb{F}_q)$ can be computed in polynomial time using methods due to Adleman, Huang and Pila (see [**4**] and [**120**]). These methods are generalizations of the method of Schoof which is used in the elliptic case. The authors are not aware of any implementation of this method for genus greater than one, since the algorithm, although easy to understand, appears very hard to implement.

In addition there is no known analogue of the improvements of Atkin and Elkies. This means that only the 'naive' Schoof algorithm is available in genus greater than one. Such an algorithm appears hopeless as a method, since the 'naive' algorithm is far too inefficient even for elliptic curves.

One can compute hyperelliptic curves using an analogue of the CM method for elliptic curves. This has been worked out in detail for the case of $g = 2$ in [**155**]. This method uses the class numbers of quartic CM fields, which are complex quadratic extensions of real quadratic fields. Analogues of the Hilbert polynomial are constructed, the zeros of which modulo p give the J-invariants of the curve. The curve is then recovered from its J-invariants.

The problem with this technique (and the reason it only applies in genus two) is that the J-invariants of a hyperelliptic curve have only been worked out for genus less than three. The invariants are the Igusa invariants [**53**] which are linked to the classical nineteenth century invariants of quintic and sextic polynomials. After the demise of classical invariant theory at the end of the century the drive to compute invariants of higher order quantics died out. Even today with the advent of computer algebra systems this seems a daunting task.

The fact that it seems unlikely that anyone can compute the order of $J_C(\mathbb{F}_q)$ for a general curve of genus four or five has led some to propose that one should not worry. For example, if I do not believe that someone can compute the order, $\#J_C(\mathbb{F}_q)$, then I do not need to worry about many of the attacks on such systems, since most attacks such as Pohlig–Hellman require knowledge of the group order. This of course means that the protocols need to be changed so that they do not require knowledge of the group order. Although this is a possible approach, it is probably to be rejected as it is assuming that a problem for which there is a *known polynomial time algorithm* will remain infeasible in the long term. This assumption is quite tenuous, even if the exponent in the polynomial complexity is relatively high. First, history shows that the asymptotics of polynomial time algorithms with high exponents often get improved, if enough research effort is invested. The Schoof algorithm for elliptic curves, and its improvements discussed in Chapter VII, provide a prime example. Second, even if the polynomial complexity remains high, comparisons in strength/key size, of the type made in Chapter I between exponential and sub-exponential complexities, apply even more forcefully for the gap between these complexities and a polynomial one. With problems available for which the best known attack is exponential or even sub-exponential, it would be hard to justify the choice of a problem with a polynomial time attack as the basis for a practical cryptosystem.

Just as for elliptic curves one could consider subfield-type hyperelliptic curves. In other words the curve C is defined over $\mathbb{F}_q$, for a small value of q, but we consider the Jacobian group $J_C(\mathbb{F}_{q^n})$ over $\mathbb{F}_{q^n}$. The advantage of this is that it is then easier to compute group orders. We shall now explain this method in detail and follow on from the discussion on zeta functions for elliptic curves in Chapter VI. We use [**66**] as a reference for this material (with a slight change of notation).

Let C be a curve of genus g defined over $\mathbb{F}_q$ and let N_n denote the number of $\mathbb{F}_{q^n}$-rational points on the curve. The zeta function for the curve is, as in Section VI.4,

$$Z(C;T) = \exp\left(\sum_{n\geq 1} \frac{N_n}{n} T^n\right).$$

For a curve of genus g, this zeta function can be shown to be of the following form:

$$Z(C;T) = \frac{P(T)}{(1-T)(1-qT)}.$$

Here, $P(T)$ is a polynomial with integer coefficients, which can be written as

$$\begin{aligned} P(T) &= 1 + a_1T + a_2T^2 + \cdots + a_{g-1}T^{g-1} + a_gT^g \\ &\quad + qa_{g-1}T^{g+1} + q^2a_{g-2}T^{g+2} + \cdots + q^{g-1}a_1T^{2g-1} + q^gT^{2g} \\ &= \prod_{i=1}^{g}(1-\alpha_iT)(1-\overline{\alpha_i}T), \end{aligned} \quad \text{(X.1)}$$

where each α_i is of absolute value $\sqrt{q}$. It then follows that

$$N_n = q^n + 1 - \sum_{i=1}^{g}(\alpha_i^n + \overline{\alpha}_i^n),\ n \geq 1.$$

The coefficients of $P(T)$ can be obtained from the power series identity [**66**]

$$\frac{Z'(T)}{Z(T)} = \sum_{i \geq 0}(N_{i+1} - q^{i+1} - 1)T^i.$$

It follows from this identity and Equation (X.1) that the values N_i, for $1 \leq i \leq g$, suffice to determine $P(T)$, and hence the α_i. Therefore, in this case, knowledge of $N_1, N_2, \ldots, N_g$ determines N_n for all $n > g$.

The fact that the roots of the polynomial $P(T)$ have magnitude $\sqrt{q}$ is referred to as the Riemann hypothesis for function fields as certain symmetries and properties of the zeta function follow from it, in common with the ordinary zeta function (see e.g. [**147**]). This generalization of the Hasse Theorem for elliptic curves was conjectured by Weil and proven by him for curves and abelian varieties. A more general version for projective varieties of dimension n, as well as the Riemann hypothesis for such varieties, was proven by Deligne (see [**147**]).

Our interest is in the order of $J_C(\mathbb{F}_{q^n})$. It can be shown that

$$\#J_C(\mathbb{F}_{q^n}) = \prod_{j=1}^{g}|1-\alpha_j^n|^2,$$

and thus the size of the Jacobian group of C defined over any extension field of $\mathbb{F}_q$ is also uniquely determined by the zeta function of the curve.

For example the curve

$$C : Y^2 + Y = X^{11} + X^5 + 1$$

of genus five defined over $\mathbb{F}_2$ has the following values of N_i:

$$N_1 = 1,\ N_2 = 9,\ N_3 = 13,\ N_4 = 17,\ N_5 = 21.$$

This means that the polynomial $Z(T)$ is given by

$$Z(T) = 32T^{10} - 32T^9 + 32T^8 - 16T^7 + 8T^6 - 4T^5 + 4T^4 - 4T^3 + 4T^2 - 2T + 1.$$

So the Jacobian of C over $\mathbb{F}_{2^{31}}$ has order

$$4567053241255021910453276306785987806821240012 9$$

with a cyclic subgroup of order

$$p = 1985675322284792134979685350776516437748365223,$$

where p is a 152-bit prime number.

X.3. The Hyperelliptic Discrete Logarithm Problem

Just as for elliptic curves, one needs to avoid analogues of the 'supersingular' and 'anomalous' curves. We do not give the details here but refer the reader to the papers of Frey and Rück [**44**] and Rück [**135**]. In addition the general methods of the BSGS and kangaroo methods apply for Jacobians of hyperelliptic curves as well. The most interesting case, from a theoretical standpoint, is when the genus is large in comparison with the characteristic as there are then conjectured sub-exponential methods.

In this section, for simplicity, attention is restricted to curves defined over fields of odd characteristic and it is assumed that $H(X) = 0$. In [**2**], Adleman, De Marrais and Huang proposed a (conjectured) sub-exponential method for the DLP in Jacobians of hyperelliptic curves of large genus. This method was based on the ideas of the function field sieve algorithm which can be used to solve the DLP in $\mathbb{F}_{2^n}$ [**1**]. The function field sieve is itself based on Pollard's number field sieve, (NFS), algorithm for factoring integers [**77**].

The method of Adleman, De Marrais and Huang appears to be only of theoretical interest as for practical systems the genus is usually chosen to be small so that the underlying group operations can be performed quickly. Recently Paulus [**117**] and Flassenberg and Paulus [**43**] have implemented a method for solving discrete logarithms for Jacobians of hyperelliptic curves. Flassenberg and Paulus did not, however, use the method of Adleman, De Marrais and Huang directly. Instead they made use of the fact that hyperelliptic curves correspond to degree two function field extensions. Then using the analogy between quadratic function fields and quadratic number fields they adapted the class group method of Hafner and McCurley [**50**] (see also [**29**]). This, combined with a sieving operation, provided a working method which could be applied to hyperelliptic curves of small genus. It should be pointed out that although Flassenberg and Paulus did not solve discrete logarithm problems, their methods are such that they can be easily extended so that they do.

X.3.1. The Number Field Sieve analogue.

The conceptually easier approach of Adleman, Huang and De Marrais is explained. This method generates random elements of the function field of the form

$$f = a(x)y + b(x),$$

with coprime $a(x), b(x) \in \mathbb{F}_q[x]$. The method then tries to factor the divisor $\mathrm{div}(f)$ over a predetermined set of prime divisors (the factor base). This produces a relation in the class group which can then be used with standard matrix techniques, as are used for solving discrete logarithms in $\mathbb{F}_q^*$, to find discrete logarithms in $J_C(\mathbb{F}_q)$. In the original presentation, the factor base is chosen to be the set of all split prime divisors of small degree in K. The small degree is the drawback to curves of small genus. For elliptic curves the factor base would essentially consist of half of the points on the curve over $\mathbb{F}_q$.

The decision as to whether an element of the required form factored over the factor base was decided, in [**2**], using the fact that in random polynomial time one can factor polynomials over finite fields. In the standard NFS, factorizations are expensive and so one replaces them by a sieving procedure. Factoring polynomials over finite fields is, on the other hand, inexpensive, so for a complexity-theoretic answer one does not need to use a sieving technique. However, in practice, a sieving technique for function fields, developed by Flassenberg and Paulus, has proved to be particularly useful.

Determining the prime divisor decomposition of the function f can be done via the following proposition, once the factorization of $b^2 - a^2F$ has been found.

PROPOSITION X.1. *Let $a(x), b(x) = \mathbb{F}_q[x]$ be coprime polynomials, let f denote the function $a(x)y + b(x)$ and set*

$$N_f = N_{K/\mathbb{F}_q[x]}(a(x)y + b(x)) = b(x)^2 - a(x)^2F(x) = \prod_{i=1}^{r} p_i(x)^{m_i},$$

where $p_i(x) \in \mathbb{F}_q[x]$ are irreducible. Then $\mathrm{div}(f)$ has only ramified or split primes in its support and

$$\mathrm{div}(f) = \sum_{i=1}^{r} m_i \mathrm{div}(p_i, r_i) - (\sum_{i=1}^{r} m_i)\infty,$$

where r_i is the unique polynomial of degree less than the degree of p_i such that

$$a(x)r_i(x) + b(x) \equiv 0 \pmod{p_i(x)} \text{ or } -a(x)r_i(x) + b(x) \equiv 0 \pmod{p_i(x)}.$$

A method is needed to find polynomials $a, b \in \mathbb{F}_q[x]$ such that the divisor of the function

$$f = ay + b$$

has support on the factor base only. Just as in the NFS, it is noticed that if an element of the factor base lies in the support of f then a congruence condition between a and b can be derived. This was described in Proposition X.1 above.

We organize a sieve in the function field case as is described in [**43**]. To every polynomial $g(x) \in \mathbb{F}_q[x]$ is associated a code given by $g(q) \in \mathbb{N}$. This is a unique integer which we use to index a sieving array, which is a two dimensional matrix indexed by the polynomial codes. Each array element is initialized at the start of the sieve to the value of

$$\deg(N_{K/\mathbb{F}_q[x]}(ay+b)) = \deg(b^2 - a^2F),$$

where a and b are the polynomials whose codes represent the row and column index of the array.

The sieve proceeds by taking every element, $P = \mathrm{div}(p, r)$, of the factor base in turn. The sieving array element is decreased by the degree of p if either

$$ar + b \equiv 0 \pmod{p}$$

or

$$-ar + b \equiv 0 \pmod{p}.$$

Every polynomial, $a_0 \pmod{p}$, is taken, in the a-direction and the polynomial $b_0 = -a_0r \pmod{p}$ is computed. The degree of p is subtracted from every array element which satisfies

$$(a, b) = (a_0 + e_1p, \pm b_0 + e_2p)$$

where e_1 and e_2 are polynomials. This can be done efficiently but care needs to be taken as to how we jump through the array. Details of how this can be done can be found in [**43**].

Polynomial arithmetic is not used to compute the jumps. This would mean that in order to deduce the next array element the current array position would have to be converted to polynomials, the polynomial addition or left shift performed, and then converted back to two polynomial codes. It is far more efficient to implement polynomial addition and left shift directly on the codes themselves. A left shift is simply a multiplication by q, while an addition can be carried out efficiently by computing a base q expansion of the codes of the polynomials which need to be added.

X.3.2. The Hafner–McCurley analogue. In the method used by Paulus and Flassenberg, which is based on the ideas of Hafner and McCurley, factorization of the element $a + by$ is replaced by attempting to factor a divisor equivalent to a given random sum of elements of the factor base.

Let $\mathcal{F}$ denote the factor base of split prime divisors. The idea, just as in the previous method, is to find relations on the elements in this factor base. A random power sum of elements in $\mathcal{F}$ is first computed,

$$D = \sum_{D_i \in \mathcal{F}} [n_i]D_i.$$

If a divisor, D', can be found which is equivalent to D and which factors over the factor base as

$$D' = \sum_{D_i \in \mathcal{F}} [m_i]D_i,$$

then we have the relation

$$\sum_{D_i \in \mathcal{F}} [n_i - m_i] D_i = \mathcal{O}.$$

Every divisor D can be represented, as operations are in an imaginary quadratic function field, as a quadratic form

$$D = (a, b, c)\ ,\ a, b, c \in \mathbb{F}_q[X],$$

of discriminant $F(X)$. Prime forms are those of the form (p, b_p, c_p) with $b_p \equiv F(X) \pmod p$, $\deg b_p < \deg p$.

The rational primes, i.e. irreducible polynomials in $\mathbb{F}_q[X]$, which lie below primes in the support of D are those polynomials which are factors of a,

$$a = \epsilon \prod_p p^{v_p},$$

where $\epsilon \in \mathbb{F}_q[X]^*$. If the prime divisors of f_p are defined by

$$f_p = (p, b_p, c_p)$$

then

$$(a, b, c) \equiv \sum_p [\epsilon_p v_p] f_p$$

with $\epsilon_p = \pm 1$ and $b \equiv \epsilon_p b_p \pmod{2p\mathbb{F}_q[X]}$ and c_p such that f_p has discriminant $F(X)$.

How is such a factorization of D' over the factor base obtained? Every divisor equivalent to D is represented by a quadratic form of the shape

$$(ax^2 + bxy + cy^2, *, *).$$

Hence we need to run through a set of $(x, y) \in \mathbb{F}_q[X] \times \mathbb{F}_q[X]$ until we obtain a polynomial $ax^2 + bxy + cy^2$ which can be factored over the polynomials lying below the prime divisors in the factor base. Clearly sieving techniques, as used above, can be applied to this problem.

The method of Hafner–McCurley has been the most successful of all methods for finding group structure, and hence discrete logarithms, of Jacobians of curves of high genus. For example a curve over $\mathbb{F}_{11}$ of genus eight may take two hours to compute the group structure using BSGS type methods, but the sieving method above requires only 17 minutes.

It should be noted that the method really requires large genus curves. The same method applied to an elliptic curve over a field of size 10^5 can take two minutes using BSGS while it would take over five hours using the above methods. The crossover point of the BSGS and the sieving methods described above seems to be around fields of order 10^g, where g is the genus. Such Jacobians will have group orders about

$$10^{g^2}.$$

Notice that for a genus five curve over a field size of 32-bits we can implement a cryptosystem without using large integer arithmetic and for which the above methods cannot be applied successfully to compute discrete logarithms. On another plus side, such Jacobians will have group orders around

$$(2^{32})^5 = 2^{160},$$

and so can be made resistant, with current computing power, to the general discrete logarithm algorithms, such as BSGS.

APPENDIX A

Curve Examples

This appendix presents examples of elliptic curves whose groups of rational points contain large prime subgroups. Section A.1 shows curves over finite fields $\mathbb{F}_q$, with $q = p$, a large prime, while Section A.2 shows curves over $\mathbb{F}_q$, with $q = 2^n$. Unless explicitly noted otherwise, the curves are 'random', in the sense that their relevant coefficients were drawn at random, with uniform probability, and the orders of their groups of rational points were determined using the point counting algorithms described in Chapter VII. In each case, a number of random curves E was generated, and the order of the group $E(\mathbb{F}_q)$ determined, until a satisfactory one was found. The probability of success in such random trials was discussed in Section VI.5.

In Section A.1, we also present some examples of curves generated with the CM method described in Chapter VIII.

All the primes listed in the examples were certified using the elliptic curve primality proving (ECPP) method [7] described in Chapter IX.

A.1. Odd Characteristic

The examples in this section describe curves over fields $\mathbb{F}_p$, where p is a large prime. The curve equations are of the form

$$E: \; Y^2 = X^3 + aX + b, \quad a, b \in \mathbb{F}_p.$$

For each curve, the values of p, a, b, and $\#E(\mathbb{F}_p)$ are listed, with elements of $\mathbb{F}_p$ shown as integers in the range $\{0, 1, \dots, p-1\}$, in decimal notation. When $\#E(\mathbb{F}_p)$ is composite, it is also shown factored as $s \cdot r$, where s is a small positive integer, and r is prime. Large integers might be broken into multiple lines, with a backslash at the end of a line indicating that the number is continued in the next line.

Examples 1–7 show 'random' curves, as described above. In these examples, the values of p are all of the form $2^k + c$, c a small positive integer, so the 'size' of a field element is self-evident. The curves in examples 8–11 were generated with the CM method. For these examples, the value $\lceil \log_2 p \rceil$ is shown (since p has no special form), as are the discriminant $-D$ and the class number h_D.

In all cases, the curve initially obtained was renormalized with a transformation of the form $a \to u^4 a$, $b \to u^6 b$, $u \neq 0$, to make the coefficient a a small integer. As discussed in Chapter III, the resulting curve is isomorphic to the original one.

EXAMPLE 1.

$$\begin{aligned} p &= 2^{130} + 169 \\ &= 1361129467683753853853498429727072845993, \\ a &= 3, \\ b &= 1043498151013573141076033119958062900890, \\ \#E(\mathbb{F}_p) &= 1361129467683753853808807784495688874237 \\ & \text{(a prime number).} \end{aligned}$$

EXAMPLE 2.

$$\begin{aligned} p &= 2^{130} + 169 \\ &= 1361129467683753853853498429727072845993, \\ a &= 1, \\ b &= 1230929586093851880935564157041535079194, \\ \#E(\mathbb{F}_p) &= 1361129467683753853846060531160085896483 \\ & \text{(a prime number).} \end{aligned}$$

EXAMPLE 3.

$$\begin{aligned} p &= 2^{160} + 7 \\ &= 1461501637330902918203684832716283019655932542983, \\ a &= 10, \\ b &= 1343632762150092499701637438970764818528075565078, \\ \#E(\mathbb{F}_p) &= 1461501637330902918203683518218126812711137002561 \\ & \text{(a prime number).} \end{aligned}$$

EXAMPLE 4.

$$\begin{aligned} p &= 2^{160} + 7 \\ &= 1461501637330902918203684832716283019655932542983, \\ a &= 1, \\ b &= 1010685925500572430206879608558642904226772615919, \\ \#E(\mathbb{F}_p) &= 1461501637330902918203683038630093524408650319587 \\ & \text{(a prime number).} \end{aligned}$$

EXAMPLE 5.

$$
\begin{aligned}
p &= 2^{190} + 129 \\
&= 156927543384667019095894735580191660402558886111600\backslash \\
&\quad 08628353, \\
a &= 10, \\
b &= 134846241141436131261105411311693108758069491867740\backslash \\
&\quad 22294274, \\
\#E(\mathbb{F}_p) &= 156927543384667019095894735578028704030525554089690\backslash \\
&\quad 46997883 \\
&\quad \text{(a prime number).}
\end{aligned}
$$

EXAMPLE 6.

$$
\begin{aligned}
p &= 2^{190} + 129 \\
&= 156927543384667019095894735580191660402558886111600\backslash \\
&\quad 08628353, \\
a &= 2, \\
b &= 123522467123718858718668331484303955154914555165230\backslash \\
&\quad 48919785, \\
\#E(\mathbb{F}_p) &= 156927543384667019095894735574486042818733937927820\backslash \\
&\quad 34198947 \\
&\quad \text{(a prime number).}
\end{aligned}
$$

EXAMPLE 7.

$$
\begin{aligned}
p &= 2^{230} + 67 \\
&= 172543658669764094685868896556925636311277724304250\backslash \\
&\quad 966387906310559498 91, \\
a &= 7, \\
b &= 307606271659321167080093083428860169417441886151220\backslash \\
&\quad 817540619633362515, \\
\#E(\mathbb{F}_p) &= 172543658669764094685868896556925634956787638464620\backslash \\
&\quad 097011905421233552 79 \\
&= 3 \cdot 575145528899213648952896321856418783189292128210\backslash \\
&\quad 540323373018070778 5093.
\end{aligned}
$$

EXAMPLE 8. CM method: $D = 120$, class number $h_D = 4$.

$$
\begin{aligned}
p &= 65455032684134289174663529622762084395449925683109\backslash \\
&\quad 5265159541652712020456264658138887199159, \\
\lceil \log_2 p \rceil &= 299, \\
a &= 1, \\
b &= 36144049423228325504814614498592213402384746101106\backslash \\
&\quad 1544601950323468977393822343563902480881, \\
\#E(\mathbb{F}_p) &= 65455032684134289174663529622762084395449925575241\backslash \\
&\quad 0858744956803878193824452885839260141966 \\
&= 2 \cdot 32727516342067144587331764811381042197724962787\backslash \\
&\quad 6205429372478401939096912226442919630070983.
\end{aligned}
$$

EXAMPLE 9. CM method: $D = 532$, class number $h_D = 4$.

$$
\begin{aligned}
p &= 75351630183031237089471027567747356575330769048527\backslash \\
&\quad 5188329021453480578403168297870032548617, \\
\lceil \log_2 p \rceil &= 299, \\
a &= 5, \\
b &= 88724011078561722486617538567703959028485726225214\backslash \\
&\quad 9486851149501578553921103683825690162, \\
\#E(\mathbb{F}_p) &= 75351630183031237089471027567747356575330769209158\backslash \\
&\quad 0197738211634229731628612489372399302238 \\
&= 2 \cdot 37675815091515618544735513783873678287665384604\backslash \\
&\quad 5790098869105817114865814306244686199651119.
\end{aligned}
$$

EXAMPLE 10. CM method: $D = 120$, class number $h_D = 4$.

$$
\begin{aligned}
p &= 2127953884222890683207317883732010798582054423945 2\backslash \\
&\quad 0064329055150059963843051285907066506563077392060 6\backslash \\
&\quad 8593671762873153882 71, \\
\lceil \log_2 p \rceil &= 400, \\
a &= 3, \\
b &= 830490090043453610772294255430609809085774865121 63\backslash \\
&\quad 203435134636821407110814962310016493084956120102 28\backslash \\
&\quad 477120368855362169 02, \\
\#E(\mathbb{F}_p) &= 2127953884222890683207317883732010798582054423945 2\backslash \\
&\quad 0064329055305636884730989494285190914590082456356 9\backslash \\
&\quad 61233459586018329307 4 \\
&= 22 \cdot 9672517655558594014578717653327321811736611017\backslash \\
&\quad 9327301967752411653129423177042856904961177310207 4\backslash \\
&\quad 349823788452663719678 67.
\end{aligned}
$$

EXAMPLE 11. CM method: $D = 307$, class number $h_D = 3$.

$$
\begin{aligned}
p &= 7048845069432712742002816418648618696753822818038 7\backslash \\
&\quad 4374287823572590636465776430902994937116627154697 5\backslash \\
&\quad 96008175843994317887, \\
\lceil \log_2 p \rceil &= 399, \\
a &= 5, \\
b &= 3866629042208848461581189787552969575881611445812 2\backslash \\
&\quad 7227632608477394833508761427897436830503346162919 4\backslash \\
&\quad 63497627079364752199, \\
\#E(\mathbb{F}_p) &= 7048845069432712742002816418648618696753822818038 7\backslash \\
&\quad 4374287823399937553496810645471164576003122183606 1\backslash \\
&\quad 60284656185776243884 \\
&= 4 \cdot 1762211267358178185500704104662154674188455704 5\backslash \\
&\quad 0968593571955849984388374202661367791144000780545 9\backslash \\
&\quad 01540071164046444060971.
\end{aligned}
$$

A.2. Characteristic Two

The examples in this section describe curves over fields $\mathbb{F}_q$ with $q = 2^n$, defined by equations of the form

$$E: \ Y^2 + XY = X^3 + a_2X^2 + a_6, \quad a_2, a_6 \in \mathbb{F}_q.$$

For each curve, the values of n, $f(x)$, a_2, a_6, and $\#E(\mathbb{F}_q)$ are listed, where $f(x)$ is the irreducible polynomial used to represent $\mathbb{F}_q$ over $\mathbb{F}_2$. The coefficient a_2 is, in all the examples, either 0 or 1 and thus equal to its trace, as all values of n listed are odd. The coefficient a_6 is presented in hexadecimal form. Each hexadecimal digit expands in the natural way to four bits, except possibly the most significant digit, which expands to the appropriate number of bits for a total length of n. Once expanded, the bits represent the coefficients of $\alpha^{n-1}, \alpha^{n-2}, \ldots, \alpha^0$, respectively from left to right, where α is a root of $f(x)$. The group order $\#E(\mathbb{F}_q)$ is shown in decimal form, and also factored as $s \cdot r$, where s is a small positive integer, and r is prime. In all the examples, s is the smallest possible value for the given isomorphism class, i.e., $s = 2$ when $Tr_2(a_2) = 1$, $s = 4$ otherwise. As before, a backslash at the end of a line indicates that the number (hexadecimal or decimal) is continued in the next line. All curves in this section are 'random'.

EXAMPLE 12. $n = 131$, $f(x) = x^{131} + x^8 + x^3 + x^2 + 1$,

$$\begin{aligned} a_2 &= 1, \\ a_6 &= \text{7417501D24550DBC77351632C8513E8FE}, \\ \#E(\mathbb{F}_q) &= 2722258935367507707729351292932711465734 \\ &= 2 \cdot 1361129467683753853864675646466355732867. \end{aligned}$$

EXAMPLE 13. $n = 131$, $f(x) = x^{131} + x^8 + x^3 + x^2 + 1$,

$$\begin{aligned} a_2 &= 0, \\ a_6 &= \text{4AC7797773F8A77E6303D3D77655D6924}, \\ \#E(\mathbb{F}_q) &= 2722258935367507707780951897749277506950 8 \\ &= 4 \cdot 680564733841876926952379744373193767377. \end{aligned}$$

EXAMPLE 14. $n = 163$, $f(x) = x^{163} + x^7 + x^6 + x^3 + 1$,

$$\begin{aligned} a_2 &= 1, \\ a_6 &= \text{15E6478546D92CE2625DB7475B43689E6E40D4AD4}, \\ \#E(\mathbb{F}_q) &= 11692013098647223345629485326803604448910923041922 \\ &= 2 \cdot 584600654932361167281474266340180222445546152 09\backslash \\ & 61. \end{aligned}$$

EXAMPLE 15. $n = 163,\ f(x) = x^{163} + x^7 + x^6 + x^3 + 1,$

$$\begin{aligned} a_2 &= 0, \\ a_6 &= \text{48419ECBC9470895FC140C851849CF6F1977FF03B}, \\ \#E(\mathbb{F}_q) &= 11692013098647223345629482613505893115770279035908 \\ &= 4 \cdot 2923003274661805836407370653376473278942569758{9}\backslash \\ &\quad 77. \end{aligned}$$

EXAMPLE 16. $n = 191,\ f(x) = x^{191} + x^9 + 1,$

$$\begin{aligned} a_2 &= 1, \\ a_6 &= \text{7BC86E2102902EC4D5890E8B6B4981FF27E0482750FE}\backslash \\ &\quad \text{FC03}, \\ \#E(\mathbb{F}_q) &= 3138550867693340381917894711669229991630522301735{5}\backslash \\ &\quad 90858398 \\ &= 2 \cdot 1569275433846670190958947355834614995815261150{8}\backslash \\ &\quad 67795429199. \end{aligned}$$

EXAMPLE 17. $n = 191,\ f(x) = x^{191} + x^9 + 1,$

$$\begin{aligned} a_2 &= 0, \\ a_6 &= \text{315BB01ABA43F480142F4E87D289C59D9754AB5200A7}\backslash \\ &\quad 489, \\ \#E(\mathbb{F}_q) &= 3138550867693340381917894711656150913519945463658{9}\backslash \\ &\quad 96854612 \\ &= 4 \cdot 7846377169233350954794736779140377283799863659{1}\backslash \\ &\quad 4749213653. \end{aligned}$$

EXAMPLE 18. $n = 239,\ f(x) = x^{239} + x^{36} + 1,$

$$\begin{aligned} a_2 &= 1, \\ a_6 &= \text{6BAB7A91D4794C8971A80A6A48B1DF53A464297EE089}\backslash \\ &\quad \text{6C2EB097D93E4F0}, \\ \#E(\mathbb{F}_q) &= 8834235323891921647916487503714592591590290962065{4}\backslash \\ &\quad 780009456530402909108 6 \\ &= 2 \cdot 4417117661945960823958243751857296295795145481{0}\backslash \\ &\quad 327390004728265201454554 3. \end{aligned}$$

EXAMPLE 19. $n = 239,\ f(x) = x^{239} + x^{36} + 1,$

$$
\begin{aligned}
a_2 &= 0,\\
a_6 &= \text{52BCEACD14FB3DCBCE421A3C6E59D4B663215EFF1457}\backslash\\
&\ \text{498E4ABB6412CFA5},\\
\#E(\mathbb{F}_q) &= 8834235323891921647916487503714592593688209043762 2\backslash\\
&\ 493128243529667661618 8\\
&= 4 \cdot 220855883097298041197912187592864814842205226 09\backslash\\
&\ 405623282060882416915404 7.
\end{aligned}
$$

EXAMPLE 20. $n = 307,\ f(x) = x^{307} + x^8 + x^4 + x^2 + 1,$

$$
\begin{aligned}
a_2 &= 1,\\
a_6 &= \text{393C7F7D53666B5054B5E6C6D3DE94F4296C0C599E2E}\backslash\\
&\ \text{2E241050DF18B6090BDC90186904968BB},\\
\#E(\mathbb{F}_q) &= 26074060497081421904236104811640040461458795438640\backslash\\
&\ 655854612651119232184598621501873866181412 6\\
&= 2 \cdot 13037030248540710952118052405820020230729397719\backslash\\
&\ 3203279273063255596160922993107509369330907063.
\end{aligned}
$$

EXAMPLE 21. $n = 367,\ f(x) = x^{367} + x^{21} + 1,$

$$
\begin{aligned}
a_2 &= 1,\\
a_6 &= \text{43FC8AD242B0B7A6F3D1627AD5654447556B47BF6AA4}\backslash\\
&\ \text{A64B0C2AFE42CADAB8F93D92394C79A79755437B5699}\backslash\\
&\ \text{5136},\\
\#E(\mathbb{F}_q) &= 30061345059505065316985351638903513950408736626026\backslash\\
&\ 49434804528587635001816368941330023634165866357513\backslash\\
&\ 18745406098\\
&= 2 \cdot 15030672529752532658492675819451756975204368313\backslash\\
&\ 01324717402264293817500908184470665011817082933178\backslash\\
&\ 75659372703049.
\end{aligned}
$$

EXAMPLE 22. $n = 401,\ f(x) = x^{401} + x^{152} + 1,$

$$\begin{aligned}
a_2 &= 1, \\
a_6 &= \text{83420635F8EA519BEC743DF9DBCA94AC950E076F90C0}\backslash \\
&\quad\ \text{7C2821262E3C180FF8A2D2F4AF6DF2FB1833EFCEE99E}\backslash \\
&\quad\ \text{811CFB11CFA0}, \\
\#E(\mathbb{F}_q) &= 516449975617381717931183834400602374865941158565584\backslash \\
&\quad\ 470256613196992421507156774502188854599840025461455\backslash \\
&\quad\ 032989725132571785934 \\
&= 2 \cdot 25822498780869085896559191720030118743297057928\backslash \\
&\quad\ 29223512830659849621075357838725109442729992001273\backslash \\
&\quad\ 072516494862566285892967.
\end{aligned}$$

EXAMPLE 23. $n = 431,\ f(x) = x^{431} + x^{120} + 1,$

$$\begin{aligned}
a_2 &= 1, \\
a_6 &= \text{715C87C2294703FF4B46C0BC257F89AE9E420BF6F07D}\backslash \\
&\quad\ \text{1E80A537F7269DAE06D7CD9EDECBCCF777D7D041F888}\backslash \\
&\quad\ \text{9D5C51A61C93DCC266CE}, \\
\#E(\mathbb{F}_q) &= 55453393882416297191568283682861674068728741507516\backslash \\
&\quad\ 33150340959161171808908283429806884365866090618516\backslash \\
&\quad\ 771699707619208876544223742366 \\
&= 2 \cdot 27726696941208148595784141841430837034364370753\backslash \\
&\quad\ 75816575170479580585904454141714903442182933045309\backslash \\
&\quad\ 258385849853809604438272111871183.
\end{aligned}$$

Bibliography

[A-1] L.M. Adleman and M.-D. Huang, editors. *ANTS-1: Algorithmic Number Theory*. Springer-Verlag, LNCS 877, 1994.

[A-2] H. Cohen, editor. *ANTS-2: Algorithmic Number Theory*. Springer-Verlag, LNCS 1122, 1996.

[A-3] J. Buhler, editor. *ANTS-3: Algorithmic Number Theory*. Springer-Verlag, LNCS 1423, 1998.

[A92] J. Seberry and Y. Zheng, editors. *Advances in Cryptology, AUSCRYPT 92*. Springer-Verlag, LNCS 718, 1993.

[A94] J. Pieprzyk and R. Safavi-Naini, editors. *Advances in Cryptology, ASIACRYPT 94*. Springer-Verlag, LNCS 917, 1995.

[A98] K. Ohta and D. Pei, editors. *Advances in Cryptology, ASIACRYPT 98*. Springer-Verlag, LNCS 1514, 1998.

[B98] D.A. Buell and J.T. Teitelbaum, editors. *Computational Perspectives on Number Theory: Proceedings of a Conference in Honor of A.O.L. Atkin*, American Mathematical Society International Press, **7**, 1998.

[C85] H.C. Williams, editor. *Advances in Cryptology, CRYPTO 85*. Springer-Verlag, LNCS 218, 1986.

[C90] A.J. Menezes and S.A. Vanstone, editors. *Advances in Cryptology, CRYPTO 90*. Springer-Verlag, LNCS 537, 1991.

[C91] J. Feigenbaum, editor. *Advances in Cryptology, CRYPTO 91*. Springer-Verlag, LNCS 576, 1992.

[C92] E.F. Bickell, editor. *Advances in Cryptology, CRYPTO 92*. Springer-Verlag, LNCS 740, 1992.

[C94] Y.G. Desmedt, editor. *Advances in Cryptology, CRYPTO 94*. Springer-Verlag, LNCS 839, 1994.

[C96] N. Koblitz, editor. *Advances in Cryptology, CRYPTO 96*. Springer-Verlag, LNCS 1109, 1996.

[C97] B. Kaliski, editor. *Advances in Cryptology, CRYPTO 97*. Springer-Verlag, LNCS 1294, 1997.

[E84] F. Pichler, editor. *Advances in Cryptology, EUROCRYPT 84*. Springer-Verlag, LNCS 219, 1985.

[E89] J.-J. Quisquater and J. Vandewalle, editors. *Advances in Cryptology, EUROCRYPT 89*. Springer-Verlag, LNCS 434, 1990.

[E90] I.B. Damgard, editor. *Advances in Cryptology, EUROCRYPT 90*. Springer-Verlag, LNCS 473, 1991.

[E91] D.W. Davies. *Advances in Cryptology, EUROCRYPT 91*. Springer-Verlag, LNCS 547, 1991.

[E95] L.C. Guillou and J.-J. Quisquater, editors. *Advances in Cryptology, EUROCRYPT 95*. Springer-Verlag, LNCS 921, 1995.

[E96] U.M. Maurer, editor. *Advances in Cryptology, EUROCRYPT 96*. Springer-Verlag, LNCS 1070, 1996.

[E97] W. Fumy, editor. *Advances in Cryptology, EUROCRYPT 97*. Springer-Verlag, LNCS 1233, 1997.

[E98] K. Nyberg, editor. *Advances in Cryptology, EUROCRYPT 98*. Springer-Verlag, LNCS 1403, 1998.

[FIPS186] FIPS 186. *Digital Signature Standard*. Federal Information Processing Standards Publication 186, U.S. Department of Commerce/N.I.S.T. National Technical Information Service, 1994.

[P1363] IEEE P1363/D3 (Draft version 3). Standard specifications for public key cryptography. May 1998.

[1] L. Adleman. The function field sieve. In [**A-1**], 108–121.

[2] L. Adleman, J. De Marrais, and M.-D. Huang. A sub-exponential algorithm for discrete logarithms over the rational subgroup of the Jacobians of large genus hyperelliptic curves over finite fields. In [**A-1**], 28–40.

[3] L. Adleman and M.-D. Huang. *Primality Testing and Abelian Varieties over Finite Fields*. Springer-Verlag, LNM 1512, 1992.

[4] L. Adleman and M.-D. Huang. Counting rational points on curves and abelian varieties over finite fields. In [**A-2**], 1–16.

[5] A.V. Aho, J.E. Hopcroft and J.D. Ullman. *The Design and Analysis of Computer Algorithms*. Addison-Wesley Publishing Co., 1974.

[6] S. Arno and F.S. Wheeler. Signed digit representations of minimal Hamming weight. *IEEE Trans. Comp.*, **42**, 1007–1010, 1993.

[7] A.O.L. Atkin and F. Morain. Elliptic curves and primality proving. *Math. Comp.*, **61**, 29–67, 1993.

[8] R. Balasubramanian and N. Koblitz. The improbability that an elliptic curve has sub-exponential discrete log problem under the Menezes–Okamoto–Vanstone algorithm. *J. Crypto.*, **11**, 141–145, 1998.

[9] E.R. Berlekamp. *Algebraic Coding Theory*. Aegean Park Press, 1984.

[10] I. Biehl, J. Buchmann and C. Thiel. Cryptographic protocols based on discrete logarithms in real-quadratic orders. In [**C94**], 56–60.

[11] B.J. Birch. Atkin and the Atlas Lab. In [**B98**], 13–20.

[12] B.J. Birch and H.P.F. Swinnerton-Dyer. Notes on elliptic curves. I. *J. Reine Angew. Math.*, **212**, 7–25, 1963.

[13] B.J. Birch and H.P.F. Swinnerton-Dyer. Notes on elliptic curves. II. *J. Reine Angew. Math.*, **218**, 79–108, 1965.

[14] I.F. Blake, S. Gao and R.J. Lambert. Construction and distribution problems for irreducible trinomials over finite fields. In *Applications of Finite Fields*, D. Gollman, editor, Oxford University Press, 1996.

[15] I.F. Blake, X.H. Gao, R.C. Mullin, S.A. Vanstone and T. Yaghoobian. *Applications of Finite Fields*. A.J. Menezes, Editor. Kluwer Academic Publishers, 1993.

[16] I.F. Blake, R.M. Roth, G. Seroussi. Efficient arithmetic in finite fields through palindromic representation. Hewlett–Packard Technical Report No. HPL-98-134, August 1998.

[17] D. Bleichenbacher. On the security of the KMOV public key cryptosystem. In [**C97**], 235–248.

[18] D. Boneh and R. Lipton. Algorithms for black-box fields and their application to cryptography. In [**C96**], 283–297.

[19] D. Boneh and R. Venkatesan. Breaking RSA may not be equivalent to factoring. In [**E98**], 59–71.

[20] J. Buchmann, S. Düllman and H.C. Williams. On the complexity and efficiency of a new key exchange system. In [**E89**], 597–616.

[21] J. Buchmann, M. Jacobson and E. Teske. On some computational problems in finite abelian groups. *Math. Comp.*, **66**, 1663–1687, 1997.

[22] J. Buchmann and S. Paulus. A one way function based on ideal arithmetic in number fields. In [**C97**], 385–394.

[23] J. Buchmann and H.C. Williams. A key-exchange system based on imaginary quadratic fields. *J. Crypto.*, **1**, 107–118, 1988.

[24] D.G. Cantor. Computing in the Jacobian of a hyperelliptic curve. *Math. Comp.*, **48**, 95–101, 1987.

[25] J.W.S. Cassels. Diophantine equations with special reference to elliptic curves. *J. LMS*, **41**, 193–291, 1966.

[26] L. S. Charlap, R. Coley and D. P. Robbins. Enumeration of rational points on elliptic curves over finite fields. *Preprint*, 1992.

[27] D.V. Chudnovsky and G.V. Chudnovsky. Sequences of numbers generated by addition in formal groups and new primality and factorization tests. *Adv. in Appl. Math.*, **7**, 385–434, 1987.

[28] W.E. Clark and J.J. Liang. On arithmetic weight for a general radix representation of integers. *IEEE Trans. Info. Theory*, **19**, 823–826, 1973.

[29] H. Cohen. *A Course In Computational Algebraic Number Theory*. Springer-Verlag, GTM 138, 1993.

[30] H. Cohen, A. Miyaji and T. Ono. Efficient elliptic curve exponentiation using mixed coordinates. In [**A98**], 51–65.

[31] P. Cohen. On the coefficients of the transformation polynomials for the elliptic modular function. *Math. Proc. Camb. Phil. Soc.*, **95**, 389–402, 1984.

[32] J.-M. Couveignes. Computing a square root for the number field sieve. In [**77**], 95–102.

[33] J.-M. Couveignes. Computing l-isogenies using the p-torsion. In [**A-2**], 59–65.

[34] J.-M. Couveignes. *Quelques calculs en théorie des nombres*. Thèse, Université de Bordeaux I, July 1994.

[35] J.-M. Couveignes. *Isogeny cycles and the Schoof–Elkies–Atkin algorithm*. L'École Polytechnique, Laboratoire D'Informatique, CNRS, Palaiseau, August, 1996.

[36] J.-M. Couveignes and F. Morain. Schoof's algorithm and isogeny cycles. In [**A-1**], 43–58.

[37] R. Crandall. Method and apparatus for public key exchange in a cryptographic system. U.S. Patent Number 5159632, 1992.

[38] S.R. Dussé and B.S. Kaliski. A cryptographic library for the Motorola DSP56000. In [**E90**], 230–244.

[39] T. ElGamal. A public key cryptosystem and a signature scheme based on discrete logarithms. *IEEE Trans. Info. Theory*, **31**, 469-472, 1985.

[40] N.D. Elkies. Elliptic and modular curves over finite fields and related computational issues. In [**B98**], 21–76.

[41] P. Erdös. Remarks on number theory. III. On addition chains. *Acta Arith.*, 77–81, 1960.

[42] W. Feller. *An Introduction to Probability Theory and its Applications*. John Wiley & Sons, 1970.

[43] R. Flassenberg and S. Paulus. Sieving in function fields. *Preprint*, 1997.

[44] G. Frey and H.-G. Rück. A remark concerning m-divisibility and the discrete logarithm problem in the divisor class group of curves. *Math. Comp.*, **62**, 865–874, 1994.

[45] S. Gao and H.W. Lenstra. Optimal normal bases. *Designs, Codes and Cryptography*, **2**, 315–323, 1992.

[46] S. Goldwasser and J. Kilian. Almost all primes can be quickly certified. In *Proc. 18th STOC*, 316–329, 1986.

[47] S.W. Golomb. *Shift Register Sequences*. Holden–Day, 1967.

[48] D.M. Gordon. A survey of fast exponentiation methods. *J. Algorithms*, **27**, 129–146, 1998.

[49] J. Guajardo and C. Paar. Efficient algorithms for elliptic curve cryptosystems. In [**C97**], 342–356.

[50] J.L. Hafner and K.S. McCurley. A rigorous sub-exponential algorithm for computation of class groups. *J. AMS*, **2**, 837–850, 1989.

[51] O. Herrman. Über die Berechnung der Fourierkoeffizienten der Funktion $j(\tau)$. *J. Reine Angew. Math.*, **274/275**, 187–195, 1975.

[52] D. Hühnlein, M. Jacobson, S. Paulus and T. Takagi. A cryptosystem based on non-maximal imaginary quadratic orders with fast decryption. In [**E98**], 279–287.

[53] J.I. Igusa. Arithmetic variety of moduli for genus two. *Ann. Math.*, **72**, 612–649, 1960.

[54] T. Itoh and S. Tsujii. A fast algorithm for computing multiplicative inverses in $GF(2^m)$ using normal bases. *Info. and Comput.*, **78**(3), 171–177, 1988.

[55] M. Jacobson, N. Koblitz, J.H. Silverman, A. Stein and E. Teske. Analysis of the Xedni calculus attack. *Preprint*, 1999.

[56] J. Jedwab and C.J. Mitchell. Minimum weight modified signed-digit representations and fast exponentiation. *Electronics Letters*, **25**, 1171–1172, 1989.

[57] M. Joye and J.-J. Quisquater. Reducing the elliptic curve cryptosystem of Meyer–Müller to the cryptosystem of Rabin–Williams. *Designs, Codes and Cryptography*, **14**, 53–56, 1998.

[58] B.S. Kaliski. The Montgomery inverse and its applications. *IEEE Trans. Comp.*, **44**, 1064–1065, 1995.

[59] A. Karatsuba. *Doklady Akad. Nauk SSSR*, **145**, 293–294, 1962. English translation in *Soviet Physics–Doklady*, **7**, 595–596, 1963.

[60] A. W. Knapp. *Elliptic Curves*. Princeton University Press, 1993.

[61] D.E. Knuth. *The Art of Computer Programming*, **2** – *Semi-numerical Algorithms*. Addison-Wesley, 2nd edition, 1981.

[62] N. Koblitz. Elliptic curve cryptosystems. *Math. Comp.*, **48**, 203–209, 1987.

[63] N. Koblitz. Hyperelliptic cryptosystems. *J. Crypto.*, **1**, 139–150, 1989.

[64] N. Koblitz. Constructing elliptic curve cryptosystems in characteristic 2. In [**C90**], 156–167.

[65] N. Koblitz. CM-curves with good cryptographic properties. In [**C91**], 279–287.

[66] N. Koblitz, *Algebraic aspects of cryptography. 3, Algorithms and Computation in Mathematics*, Springer-Verlag, Berlin, 1998.

[67] C.K. Koç and T. Acar. Montgomery multiplication in $GF(2^k)$. *Designs, Codes and Cryptography*, **14**, 57–69, 1998.

[68] K. Koyama, U. Maurer, T. Okamoto and S.A. Vanstone. New public-key scheme based on elliptic curves over the ring $\mathbb{Z}_n$. In [**C91**], 252–266.

[69] K. Koyama and Y. Tsuruoka. Speeding up elliptic cryptosystems by using a signed binary window method. In [**C92**], 345–357.

[70] K. Kurosawa, K. Okada and S. Tsujii. Low exponent attack against elliptic curve RSA. In [**A94**], 376–383.

[71] K.-Y. Lam and L.C.K. Hui. Efficiency of SS(l) square-and-multiply exponentiation algorithms. *Electronics Letters*, **30**, 2115–2116, 1994.

[72] S. Lang. *Elliptic Curves: Diophantine Analysis*. Springer-Verlag, 1978.

[73] G.-J. Lay and H.G. Zimmer. Constructing elliptic curves with given group order over large finite fields. In [**A-1**], 250–263.

[74] F. Lehmann, M. Maurer, V. Müller and V. Shoup. Counting the number of points on elliptic curves over finite fields of characteristic greater than three. In [**A-1**], 60–70.

[75] F. Lemmermeyer. The Euclidean algorithm in algebraic number fields. *Expo. Math.*, **13**, 385–416, 1995.

[76] A. Lempel, G. Seroussi, and S. Winograd. On the complexity of multiplication in finite fields. *Theoretical Comp. Sci.*, **22**, 285–296, 1983.

[77] A.K. Lenstra and H.W. Lenstra, editors. *The Development of the Number Field Sieve.* Springer-Verlag, LNM 1554, 1993.

[78] H.W. Lenstra. Factoring integers with elliptic curves. *Ann. Math.*, **126**, 649–673, 1987.

[79] H.W. Lenstra and C.P. Schnorr. A Monte Carlo factoring algorithm with linear storage. *Math. Comp.*, **43**, 289–311, 1984.

[80] R. Lercier. Computing isogenies in $\mathbb{F}_{2^n}$. In [**A-2**], 197–212.

[81] R. Lercier. *Algorithmique des courbes elliptiques dans les corps finis.* Thèse, L'École Polytechnique, Laboratoire D'Informatique, CNRS, Paris, June, 1997.

[82] R. Lercier. Finding good random elliptic curves for cryptosystems defined over $\mathbb{F}_{2^n}$. In [**E97**], 379–392.

[83] R. Lercier and F. Morain. Counting the number of points on elliptic curves over finite fields: strategies and performances. In [**E95**], 79–94.

[84] R. Lercier and F. Morain. Algorithms for computing isogenies between elliptic curves. In [**B98**], 77-96.

[85] R. Lercier and F. Morain. Counting points on elliptic curves over $\mathbb{F}_{p^n}$ using Couveignes algorithm. Rapport de Recherche LIX/RR/95/09, 1995.

[86] R. Lidl and H. Niederreiter. *Finite Fields,* in *Encyclopedia of Mathematics and its Applications,* G.-C. Rota, editor, Addison-Wesley, 1983.

[87] J.H. van Lint. *Introduction to Coding Theory.* Springer-Verlag, 1982.

[88] K.S. McCurley. The discrete logarithm problem. In *Cryptology and Computational Number Theory,* C. Pomerance, editor, 49–74. Proc. Symp. Applied Maths **42**, 1990.

[89] J. McKee. Subtleties in the distribution of the numbers of points on elliptic curves over a finite prime field. *J. LMS,* **59**, 448–460, 1999.

[90] J. McKee and R.G.E. Pinch. On a cryptosystem of Vanstone and Zuccherato. *Preprint,* 1998.

[91] K. Mahler. On the coefficients of the 2^mth transformation polynomial for $j(\omega)$. *Acta Arith.,* **21**, 89–97, 1972.

[92] K. Mahler. On the coefficients of transformation polynomials for the modular functions. *Bull. Austral. Math. Soc.*, **10**, 197–218, 1974.

[93] J.L. Massey and O.N. Garcia. Error correcting codes in computer arithmetic. In *Advances in Information Systems Science,* J.L. Tou, editor, **4**, 273–326. Plenum, New York, 1971.

[94] U.M. Maurer. Towards the equivalence of breaking the Diffie–Hellman protocol and computing discrete logarithms. In [**C94**], 271–281.

[95] U.M. Maurer and S. Wolf. Diffie–Hellman oracles. In [**C96**], 268–282.

[96] W. Meier and O. Staffelbach. Efficient multiplication on certain non-supersingular elliptic curves. In [**C92**], 333–344.

[97] A.J. Menezes. *Elliptic Curve Public Key Cryptosystems.* Kluwer Academic Publishers, 1993.

[98] A.J. Menezes, T. Okamoto and S.A. Vanstone. Reducing elliptic curve logarithms to a finite field. *IEEE Trans. Info. Theory,* **39**, 1639–1646, 1993.

[99] A.J. Menezes, P.C. van Oorschot and S.A. Vanstone. *Handbook of Applied Cryptography*. CRC Press, 1996.

[100] A.J. Menezes, S.A. Vanstone and R. J. Zuccherato. Counting points on elliptic curves over $\mathbb{F}_{2^n}$ *Math. Comp.*, **60**, 407–420, 1993.

[101] B. Meyer and V. Müller. A public key cryptosystem based on elliptic curves over $\mathbb{Z}/n\mathbb{Z}$ equivalent to factoring. In [**E96**], 49–59.

[102] G. Miller. Riemann's hypothesis and test for primality. *J. Comp. and Sys. Sci.*, **13**, 300–317, 1976.

[103] V. Miller. Use of elliptic curves in cryptography. In [**C85**], 417–426.

[104] A. Miyaji. Elliptic curves over $\mathbb{F}_p$ suitable for cryptosystems. In [**A92**], 479–491.

[105] P.L. Montgomery. Modular multiplication without trial division. *Math. Comp.*, **44**, 519–521, 1985.

[106] P.L. Montgomery. Speeding the Pollard and elliptic curve methods of factorization. *Math. Comp.*, **48**, 243–264, 1987.

[107] F. Morain. Building cyclic elliptic curves modulo large primes. In [**E91**], 328–336.

[108] F. Morain. Calcul du nombre de points sur une courbe elliptique dans un corps fini: aspects algorithmiques. *J. Théorie des Nombres de Bordeaux*, **7**, 255–282, 1995.

[109] F. Morain and J. Olivos. Speeding up the computations on an elliptic curve using addition–subtraction chains. *Info. Theory Appl.*, **24**, 531–543, 1990.

[110] V. Müller. Ein Algorithmus zur Bestimmung der Punktzahl elliptischer Kurven über endlichen Körpern der Charakteristik grösser drei. Ph.D. Thesis, Universität des Saarlandes, 1995.

[111] V. Müller. Fast multiplication on elliptic curves over small fields of characteristic two. *J. Crypto.*, **11**, 219–234, 1998.

[112] R. Mullin, I. Onyszchuk, S.A. Vanstone and R. Wilson. Optimal normal bases in $GF(p^n)$. *Discrete Appl. Math.*, **22**, 149–161, 1988/89.

[113] K. Nyberg and R.A. Rueppel. Message recovery for signature schemes based on the discrete logarithm problem. *Designs, Codes and Cryptography*, **7**, 61–81, 1996.

[114] A.M. Odlyzko. Discrete logarithms in finite fields and their cryptographic significance. In [**E84**], 417–426.

[115] J. Omura and J. Massey. Computational method and apparatus for finite field arithmetic. U.S. Patent number 4,587,627, May 1986.

[116] P.C. van Oorschot and M.J. Wiener. Parallel collision search with cryptanalytic applications. *J. Crypto.*, **12**, 1–28, 1999.

[117] S. Paulus. An algorithm of sub-exponential type computing the class group of quadratic orders over principal ideal domains. In [**A-2**], 243–257.

[118] S. Paulus and H.-G. Rück. Real and imaginary quadratic representation of hyperelliptic function fields. *Math. Comp.*, **68**, 1233–1241, 1999.

[119] S. Paulus and A. Stein. Comparing real and imaginary arithmetics for divisor class groups of hyperelliptic curves. In [**A-3**], 576–591.

[120] J. Pila. Frobenius maps of abelian varieties and finding roots of unity in finite fields. *Math. Comp.*, **55**, 745–763, 1996.

[121] R.G.E. Pinch. Extending the Wiener attack to RSA-type cryptosystems. *Electronics Letters*, **31**, 1736–1738, 1995.

[122] J.-M. Piveteau. New signature scheme with message recovery. *Electronics Letters*, **29**, 2185, 1993.

[123] H.C. Pocklington. The determination of the prime and composite nature of large numbers by Fermat's theorem. *Proc. Camb. Phil. Soc.*, **18**, 29–30, 1914/16.

[124] G.C. Pohlig and M.E. Hellman. An improved algorithm for computing logarithms over $GF(p)$ and its cryptographic significance. *IEEE Trans. Info. Theory*, **24**, 106–110, 1978.

[125] J.M. Pollard. Monte Carlo methods for index computation (mod p). *Math. Comp.*, **32**, 918–924, 1978.

[126] K.C. Posch and R. Posch. Modulo reduction in residue number systems. *IEEE Trans. Parallel and Dist. Systems*, **6**, 449–454, 1995.

[127] K.C. Posch and R. Posch. Division in residue number systems involving length indicators. *J. Comp. Appl. Maths.*, **66**, 411-419, 1996.

[128] J.-J. Quisquater and J.-P. Delescaille. How easy is collision search? Application to DES. In [**E89**], 408–413.

[129] M. Rabin. Digitized signatures and public key functions as intractable as factorization. *MIT/LCS/TR-212*, MIT Laboratory for Computer Science, 1979.

[130] M. Rabin. Probabilistic algorithms for testing primality. *J. Number Theory*, **12**, 128–138, 1980.

[131] G. Reitwiesner. Binary arithmetic. *Adv. in Comp.*, **1**, 231–308, 1960.

[132] H. Riesel. *Prime Numbers and Computer Methods for Factorization.* Birkhäuser, 1985.

[133] R.L. Rivest, Shamir A. and L.M. Adleman. A method for obtaining digital signatures and public-key cryptosystems. *Comm. ACM*, **21**, 120–126, 1978.

[134] R.L. Rivest, Shamir A. and L.M. Adleman. Cryptographic communications system and method. US Patent No 4405829, 1983.

[135] H.-G. Rück. On the discrete logarithm problem in the divisor class group of curves. *Math. Comp.*, **68**, 805–806, 1999.

[136] T. Satoh and K. Araki. Fermat quotients and the polynomial time discrete log algorithm for anomalous elliptic curves. *Comm. Math. Univ. Sancti Pauli*, **47**, 81–92, 1998.

[137] J. Sattler and C.P. Schnorr. Generating random walks in groups. *Ann. Univ. Sci. Budapest. Sect. Comp.*, **6**, 65–79, 1985.

[138] E.F. Schaefer. Computing a Selmer group of a Jacobian using functions on the curve. *Math. Ann.*, **310**, 447-471, 1998.

[139] B. Schneier. *Applied Cryptography.* John Wiley and Sons, 1996.

[140] A. Schönhage. Schnelle Multiplikation von Polynomen über Körpen der Charakteristik 2. *Acta Info.*, **7**, 395–398, 1977.

[141] R. Schoof. Elliptic curves over finite fields and the computation of square roots mod p. *Math. Comp.*, **44**, 483–494, 1985.

[142] R. Schoof. Counting points on elliptic curves over finite fields. *J. Théorie des Nombres de Bordeaux*, **7**, 219–254, 1995.

[143] I.A. Semaev. Evaluation of discrete logarithms on some elliptic curves. *Math. Comp.*, **67**, 353–356, 1998.

[144] G. Seroussi. Table of low-weight irreducible polynomials over $\mathbb{F}_2$. Hewlett–Packard Laboratories Technical Report No. HPL-98-135, August 1998.

[145] G. Seroussi. Compact representation of elliptic curve points over $\mathbb{F}_{2^n}$. Hewlett–Packard Laboratories Technical Report No. HPL-98-94R1, September 1998.

[146] V. Shoup. Lower bounds for discrete logarithm and related problems. In [**E97**], 313–328.

[147] J.H. Silverman. *The Arithmetic of Elliptic Curves.* Springer-Verlag, GTM 106, 1986.

[148] J.H. Silverman. *Advanced Topics in the Arithmetic of Elliptic Curves.* Springer-Verlag, GTM 151, 1994.

[149] J.H. Silverman and J. Suzuki. Elliptic curve discrete logarithms and the index calculus. In [**A98**], 110–125.

[150] J.H. Silverman. The xedni calculus and the elliptic curve discrete logarithm problem. *Preprint*, 1998.

[151] N.P. Smart. *The Algorithmic Resolution of Diophantine Equations*. Cambridge University Press, 1998.

[152] N.P. Smart. Elliptic curves over small fields of odd characteristic *J. Crypto.*, **12**, 141–151, 1999.

[153] N.P. Smart. The discrete logarithm problem on elliptic curves of trace one. *J. Crypto.*, **12**, 193–196, 1999.

[154] J.A. Solinas. An improved algorithm for arithmetic on a family of elliptic curves. In [**C97**], 357-371.

[155] A.-M. Spallek. Kurven vom Geschlecht 2 und ihre Anwendung in Public-Key-Kryptosytemen *Ph.D. Thesis*, Universität Essen, 1994.

[156] R.G. Swan. Factorization of polynomials over finite fields. *Pacific J. Math.*, **12**, 1099–1106, 1962.

[157] E. Teske. Speeding up Pollard's Rho method for computing discrete logarithms. In [**A-3**], 541–554.

[158] E. Teske. A space efficient algorithm for group structure computation. *Math. Comp.*, **67**, 1637–1663, 1998.

[159] S.A. Vanstone and R.J. Zuccherato. Elliptic curve cryptosystems using curves of smooth order over the ring $\mathbb{Z}_n$. *IEEE Trans. Info. Theory*, **43**, 1231–1237, 1997.

[160] J. Vélu. Isogénies entre courbes elliptiques. *Comptes Rendus l'Acad. Sci. Paris, Ser. A*, **273**, 238-241 1971.

[161] J.F. Voloch. The discrete logarithm problem on elliptic curves and descents *Preprint*, 1997.

[162] A. Wiles. Modular elliptic curves and Fermat's Last Theorem. *Ann. Math.*, **142**, 443–551, 1995.

[163] H.C. Williams. A modification of the RSA public-key encryption procedure. *IEEE Trans. Info. Theory*, **26**, 726–729, 1980.

[164] S. Winograd. Some bilinear forms whose complexity depends on the field of constants. *Math. Sys. Theory*, **10**, 169–180, 1977.

Author Index

Adleman, L.M., 7, 165, 173, 176
Araki, K., 88, 106
Atkin, A.O.L., 50, 114, 116, 118, 119, 146, 156, 165, 173

Birch, B.J., 98, 114
Blake, I.F., 24
Boneh, D., x, 6, 166
Buchmann, J., x, 7

Cohen, H., 102
Cremona, J., x

De Marrais, J., 176
Deligne, P., 49
Dussé, S.R., 16

Elkies, N., 50, 114, 115, 118, 119, 139, 146, 173
Erdös, P., 65

Flassenberg, R., 176, 177
Frey, G., 82, 86, 176

Galbraith, S., x
Goldwasser, S., 164, 165

Hafner, J.L., 176
Hellman, M.E., 7, 80
Huang, M.-D., 165, 173, 176

Itoh, T., 22

Joye, M, 8

Kaliski, B.S., 16, 17
Karatsuba, A., 20
Kilian, J., 164, 165
Knuth, D.E., 63
Koblitz, N., ix, 7, 107, 171
Koyama, K., 8

Lenstra, H.W. Jnr, 35, 159
Lercier, R., 118, 133–138

Massey, J., 22
Maurer, M., x
Maurer, U.M., 166, 168
McCurley, K.S., 7, 176
Menezes, A.J., x, 2, 82, 86
Mestre, J.-F., 102, 104
Meyer, B., 8
Miller, V., ix, 7
Miyaji, A., 79, 86
Montgomery, P.L., 15
Morain, F., 156, 165
Mordell, L.J., 49
Müller, V., x, 8, 121, 131, 133, 142

Nyberg, K., 5

Okamoto, T., 82
Omura, J., 22
van Oorschot, P., 2, 96

Paterson, K., x
Paulus, S., 176, 177
Pila, J., 173
Piveteau, J.-M, 5
Pohlig, G.C., 7, 80
Pollard, J.M., 80, 93, 104, 176

Quisquater, J.-J., 8

Rabin, M., 8
Ramanujan, S., 49
Rubinstein, M., x
Rück, H.-G., 82, 86, 176
Rueppel, R.A., 5

Satoh, T., 88, 106
Scheafer, E., x
Schneier, B., 2
Schönhage, A., 20
Schoof, R., x, 104, 109, 118, 120, 127, 165, 173
Semaev, I.A., 79, 88, 106
Seroussi, G., 77

Shanks, D., 79, 92, 102, 168
Silverman, J.H., 88
Smart, N.P., 88, 106
Solinas, J., 76
Swan, R.G., 19
Swinnerton-Dyer, H.P.F., 98

Tsujii, S., 22

Vanstone, S.A., 2, 82
Vélu, J., 134
Voloch, J.F., 82

Wiener, M.J., 96
Williams, H.C., 7, 8
Wolf, S., 166, 168

Zaba, S., x

Subject Index

addition chain, 62
addition–subtraction chains, 63
affine coordinates, 30, 57–58, 60–61
anomalous curves, 79, 86, 88–91
Atkin primes, 116, 118–122, 140–142
authenticity, 1

baby step/giant step algorithm, 79, 91–93, 104, 142–144, 168, 176
Barrett reduction, 14–15
Bernoulli number, 49
Birch–Swinnerton-Dyer conjecture, 98
bit-serial multipliers, 22
BSGS, *see* baby step/giant step algorithm

Cantor's algorithm, 172
certificate of primality, 163
Chinese Remainder Theorem, 13, 80, 109, 142, 145, 159, 160
chord–tangent process, 32
class group, 7, 92, 150, 176
class number, 151, 157, 173
CM, *see* complex multiplication
complex multiplication, 46, 149–157, 162, 165, 173
confidentiality, 1
Cornacchia's algorithm, 152, 157
CRT, *see* Chinese Remainder Theorem

Data Encryption Standard, 3
Dedekind's η-function, 49, 53, 156
DES, *see* Data Encryption Standard
descent via isogeny, 82–83
DHP, *see* Diffie–Hellman problem
Diffie–Hellman key exchange, ix, 3, 6
Diffie–Hellman problem, 3, 166–169
digital signature, 2
  ElGamal, ix, 3
  Nyberg–Rueppel, 5
  with message recovery, 5
Digital Signature Algorithm, ix, 4
diophantine equation, 152, 157
discrete logarithm problem, 2, 3, 6, 7, 79–99, 166–169
  anomalous curves, 79, 88–91
  baby step/giant step algorithm, 79, 91–93
  elliptic curve, 57, 79–99
  hyperelliptic, 176–180
  index calculus methods, 97–98
  MOV attack, 79, 82–88
  Pohlig–Hellman, 79–82
  rho, lambda and kangaroo methods, 79, 93–97
discriminant, 114, 119, 150–152, 156, 157, 165, 179
division polynomials, 39–42, 115, 135
divisor, 85, 177–179
  semi-reduced, 172
divisor class group, 85
DLP, *see* discrete logarithm problem
DSA, *see* Digital Signature Algorithm
dual isogeny, 45

ECDLP, *see* elliptic curve, discrete logarithm problem
ECM, *see* elliptic curve, factoring method
ECPP, *see* elliptic curve, primality proving method
Eisenstein series, 49, 124
ElGamal digital signature, ix, 3
ElGamal encryption, ix, 3
Elkies primes, 115–116, 118–140
elliptic curve
  admissible change of variable, 31
  applications, 159–169
  checking group order, 103–104
  determining a random point, 35
  determining group order, 101–107, 109–148
  discrete logarithm problem, 7, 57, 79–99

discriminant, 30, 124, 134
efficient implementation, 57–76
endomorphism ring, 45, 149
examples, 181–189
characteristic two, 186–189
odd characteristic, 181–185
factoring method, 7, 159–162
generating with CM, 151–157
group law, 31–34
isomorphism, 31, 36–38, 47
j-invariant, 31, 47, 116, 120, 121, 123, 126, 134, 149, 153
non-singular, 30, 31, 134
over a finite field, 34–38
point addition, 31–34, 57–62
point at infinity ($\mathcal{O}$), 30
point doubling, *see* point addition, 32
point multiplication, 62–76
primality proving method, 7, 164–166
torsion structure, 42
elliptic function, 29
elliptic integrals, 29
elliptic logarithm, p-adic, 79
endomorphism, 34, 44
Euclidean algorithm, 13, 16, 17, 21, 24, 25
Euclidean domain, 76

factor base, 178
factoring, 92
Fermat's Last Theorem, 29
finite field arithmetic, 11–27
characteristic two, 19–27
normal bases, 22–25
palindromic polynomial, 24
palindromic representation, 24
polynomial bases, 19–22
solving quadratic equations, 26
subfield bases, 25–26
odd characteristic, 11–19
Barrett reduction, 14–15
moduli of special form, 12
Montgomery arithmetic, 15–17
residue number system, 13–14
solving quadratic equations, 17–19
square roots, 17–19
formal group, 89
Frobenius endomorphism, 34, 73, 110, 116, 118, 121, 140, 145
characteristic polynomial, 115, 119, 121
Frobenius expansion, 73–76
Frobenius map, *see* Frobenius endomorphism
Frobenius, trace of, x, 34, 46, 73, 79, 90, 105, 140, 153
function field, 176
function field sieve, 176

Galois cohomology, 82
Galois group, 46
Galois representation, 46
Goldwasser–Kilian primality test, 164
group exponentiation, 2, 62, 63

Hafner–McCurley method, 178–180
half-trace, 26
Hasse's Theorem, 34, 73, 77, 102, 107
Hensel's Lemma, 90
Hilbert class field, 150, 152, 155–157
Hilbert polynomial, 150, 152–155, 157, 173
Hilbert's Theorem 90, 83
hyperelliptic cryptosystems, 171–180
hyperelliptic curve, 171
arithmetic, 171–173
Jacobian, 8, 165, 171
J-invariant, 173

Igusa invariants, 174
imaginary quadratic number field, 149
imaginary quadratic orders, 7
index calculus methods, 97–98
integrity, 1
isogeny, 44, 115, 127, 134
computing
characteristic two, 133–138
odd characteristic, 122–133
degree, 44
kernel, 121, 123, 125, 127, 128, 134

Jacobi's formula, 49, 124
Jacobian, 7, 171, 176
Jacobian representation, 58
j-invariant, *see* elliptic curve, j-invariant

kangaroos, 176
Koblitz curves, 101
Kronecker congruence relation, 51

lambda method, 80, 92, 95
lattice, 46, 50, 127
Laurent series, 89
Legendre symbol, 18, 102, 120

Massey–Omura encryption, ix, 5
Massey–Omura multiplier, 22
Miller–Rabin test, 162
modular arithmetic, 11–19
  polynomial, 19–22, 24
modular function, 47
modular inversion, 13, 16
  polynomial, 21
modular multiplication, 12
  polynomial, 19
modular polynomials, 50–55, 116, 118–122, 145
  variants, 52
modular reduction, 12
  polynomial, 19
moduli of special form, 12
Montgomery arithmetic, 15–17
Montgomery multiplication, 17
Montgomery reduction, 15–16
Mordell–Weil Theorem, 98
morphism, 44
MOV attack, 82–88
MOV condition, 99
multiplication-by-m map ($[m]$), 34

NAF, *see* non-adjacent form
Neron–Tate height, 98
Newton–Raphson iteration, 89
non-adjacent form, 67
non-repudiation, 1
number field sieve, 7, 176
Nyberg–Rueppel digital signature, 5

ONB, *see* optimal normal bases
optimal normal bases, 22–25

p-adic numbers, 88
palindromic polynomial, 24
palindromic representation, 24
Pocklington–Lehmer primality test, 162–165
Pohlig–Hellman, 80–83, 91, 168, 174
point
  addition, 31–34, 57–62
    affine coordinates, 57–58, 60–61
    cost summary, 60, 62
    projective coordinates, 59–62
  at infinity ($\mathcal{O}$), 30
  compression, 76–78
  counting, x, 42, 50, 52, 101–107, 109–148, 181
  doubling, *see* point addition, 32
  multiplication, 57, 62–76
    and exponentiation, 63
    binary method, 63
    example of costs, 72
    m-ary method, 64
    modified m-ary method, 64
    of fixed point, 73
    precomputation, 64–66, 70, 73
    relative costs, 72
    signed m-ary window, 70
    signed digit method, 67
    sliding window method, 66
    window methods, 66
    with non-adjacent form representation, 68
  rational, 30
polynomial multiplication, 20
Prime Number Theorem, 107
projective coordinates, 22, 30, 58–62
  weighted, 58
proof of primality, 162
  down run, 163
public key cryptography, 1

Ramanujan τ-function, 48
random walk, 95
rational point, 30
residue number system, 13–14
rho method, 80, 92, 96
RSA, 6, 8

Schoof's algorithm, 50, 109–148, 155, 165, 173
SEA algorithm, 116
Shanks and Mestre algorithm, 104
smooth number, 159
solving quadratic equations
  characteristic two, 26
  odd characteristic, 17–19
subfield bases, 25–26
subfield curves, 73, 101, 104–106, 174
supersingular curve, 37, 45, 83

tame and wild kangaroos, 80
Tate module, 46
torsion group, 40
torsion points, 40–44, 120
  group structure, 42, 121
trace of Frobenius, *see* Frobenius, trace of
twist, 37, 38, 104, 107, 109, 146

Weber polynomials, 155–156
Weierstrass equation, 30, 123, 127
Weierstrass form, 29
Weierstrass ℘-function, 29, 47, 127
Weil pairing, 42–45, 79, 84

zeta function, 105, 174